TOPICS IN THE FORMAL METHODOLOGY OF EMPIRICAL SCIENCES

SYNTHESE LIBRARY

STUDIES IN EPISTEMOLOGY,

LOGIC, METHODOLOGY, AND PHILOSOPHY OF SCIENCE

Managing Editor:

JAAKKO HINTIKKA, *Florida State University*

Editors:

ROBERT S. COHEN, *Boston University*

DONALD DAVIDSON, *University of Chicago*

GABRIËL NUCHELMANS, *University of Leyden*

WESLEY C. SALMON, *University of Arizona*

VOLUME 135

RYSZARD WÓJCICKI

TOPICS IN THE FORMAL METHODOLOGY OF EMPIRICAL SCIENCES

D. REIDEL PUBLISHING COMPANY

DORDRECHT : HOLLAND/BOSTON : U.S.A.
LONDON : ENGLAND

Library of Congress Cataloging in Publication Data

Wójcicki, Ryszard,

Topics in the formal methodology of empirical sciences.

(Synthese library; v. 135)
Translation of Metodologia formalna nauk empirycznych.
Includes bibliography and indexes.
1. Science–Methodology. 2. Science–Philosophy.
I. Title.
Q175.W69913 501'.8 79-16327
ISBN-13: 978-94-009-8946-7 e-ISBN-13: 978-94-009-8944-3

DOI: 10.1007/978-94-009-8944-3

Distributors for Albania, Bulgaria, Chinese People's Republic, Cuba, Czechoslovakia, German Democratic Republic, Hungary, Korean People's Democratic Republic, Mongolia, Poland, Rumania, Vietnam, the U.S.S.R. and Yugoslavia

ARS POLONA
Krakowskie Przedmieście 7, 00-068 Warszawa 1, Poland

Distributors for the U.S.A., Canada and Mexico
D. REIDEL PUBLISHING COMPANY, INC.
Lincoln Building, 160 Old Derby Street, Hingham, Mass. 02043, U.S.A.

Distributors for all other countries
D. REIDEL PUBLISHING COMPANY
P.O. Box 17, 3300AA Dordrecht, Holland

Translated from Polish by Ewa Jansen

Softcover reprint of the hardcover 1st edition 1979

... philosophy is written in the great book of nature which is continually open before our eyes, but which no one can read unless he has mastered the code in which it is composed, that is, the mathematical figures and the necessary relations between them.

GALILEO

TABLE OF CONTENTS

PREFACE TO THE ENGLISH EDITION

The main objective of the considerations carried out in the book is the relations between empirical theories and phenomena described by those theories. The simplest model of relations of that sort can be obtained under the assumption that an empirical theory may be viewed as a deductive system formalized in the first order predicate calculus, whereas all phenomena examined by the theory form a relational system being an intended interpretation for the language of the theory. Soundness of the theory depends on whether its theorems are true or not under the intended interpretation.

Inadequacy of this very simple model to most of the theories dealt with in science does not consist only or even mainly in the fact that very few theories can be treated as first order theories. Much more significant shortcoming is the assumption that an empirical theory may be viewed as a theory of exactly one system – the relational structure being the intended interpretation for the language of the theory. As a rule, any phenomenon described by a certain empirical theory (course of a given chemical reaction, animals' migrations, mechanical motion of physical bodies in given conditions, reaction of a human organism to given physical stimuli, tectonical movements of the earth – the examples may be long cited) may be viewed as a set of its particular cases (a definite case of a chemical reaction of a given type, a definite case of animals' migration, and so on, and so forth). Each of such singular phenomena determines a distinct interpretation for the theory. Thus one can speak about a set of intended interpretations linked in some way rather than about a singular interpretation of this kind. Empirical theories are multireferential.

It is obvious that the relations which a theory bears to the phenomena to which it applies may be examined with formal means only under the condition that the phenomena under consideration are represented in a form of certain abstract (mathematical) systems. If we resign from the assumption that languages of empirical theories are of first order, we have to revise the idea of representing singular empirical phenomena as relational structures. The question then arises what kinds of systems can play the role of models for empirical phenomena and what criteria can decide that in a given empirical theory systems of a certain kind provide a better representation for the phenomena under examination than some others. Undertaking the problem one must go beyond the purely semantic account of empirical theories.

Properties of empirical systems are determined not only by the theory of these systems but also by individual data gathered with the help of suitable testing procedures. Hence, one of the fundamental problems of methodology of empirical sciences is to explain how the nature of testing procedures applied in a given theory may justify the decision to represent empirical phenomena by means of systems of that and not some other kind. The requirement that empirical systems should be defined so that empirical data can be treated as the data referring directly to them seems quite natural. On the other hand such a requirement need not be imposed on interpretations of the language of a theory. Making the notion of an empirical system dependent on that of testing procedures introduces an operational aspect into considerations, although it not necessarily must be identical with acceptance of the idea of operationalism, especially in its orthodox form.

Having accepted the principle that empirical systems serving as models for empirical phenomena examined may be different from those serving as interpretations of the language of the theory one resigns from the assumption often treated as fundamental for the semantic account of science. One of the direct consequences of such a step is the necessity of reconsidering the definition of soundness of a theory.

The remarks above were intended to present briefly the problems discussed in the book. Closing the presentation I wish to comment on to what extent the present edition differs from its Polish original. Modifications which I introduced are radical and often far going. In my opinion the Polish text had two main drawbacks. It was overloaded with informal considerations and at the same time formal concepts included in some parts of the book were presented in a too complicated way. Of course one of the motives to revise it was also the fact that much time has passed since I finished writing the Polish version and obviously certain decisions and ideas contained in the first edition seem not quite relevant now. So it is not only the desire to make the exposition clearer but also the reasons of substantial nature which motivated writing a revised version.

I do not think it desirable to bother the reader with a detailed discussion of all changes to which the Polish version was subjected and that is why I will confine myself to pointing out only the most significant ones.

Explanations concerning logical and set-theoretical notions applied in the book have been shortened as much as possible, in the Polish version one whole chapter was devoted to the discussion of them.

In the present edition the discussion of the notion of a regularity, which in the Polish version was carried out with reference to so-called "generic systems", concerns mainly "elementary systems", which makes it much simpler without any significant loss of generality, I believe. I accepted the assumption that proof procedures used in empirical theories are sound, i.e. if premises are true, the thesis proved on the grounds of them is also true. Of course, such an assumption simplifies the real state of affairs, one can easily quote examples of the procedures used in empirical sciences which are not sound. But this assumption simplifies not only the real state of affairs but also the considerations carried out (for instance, many definitions become much simpler). Quite obviously, simplicity of considerations could not be a sufficient justification, if not the fact that complications resulting from admitting into considerations such theories whose procedures are not sound are trivial from the formal standpoint and thus do not lead to any more difficult problems to be solved.

Two other specially important changes introduced in the present version are: bringing into prominence the notion of an empirical system (the notion is discussed in a separate chapter having no counterpart in the original version) and resignation from the term "approximate truth". The notion of approximate truth has been replaced by that of soundness of a theory.

In spite of all those changes and modifications, fundamental ideas (especially philosophical assumptions) of both the versions are identical. Also for this reason I have decided to include Preface to the Polish version, although it is not a fully adequate introduction to the present version of the monograph.

A great part of the new version of the book was prepared during my stay in Canada as a visiting scholar of Queen's University Kingston in the Academic year 1975/76. I am grateful to the University and the National Research Council of Canada for a research grant that made the conception completion of this work possible.

In conclusion I would like to acknowledge gratefully my many debts to all people whose criticism and advice allowed me to improve the book. I am especially grateful to Professor Maria Lutman for her many valuable and incisive suggestions and Dr Józef Misiek for his critical remark on the Polish version of the book.

Perhaps not so evident but very singnificant is my debt to dr Robin Giles from Queen's University, although I do not share his radically operationalistic attitude, no dobt I have been influenced by his philosophical ideas. I treasure very much discussions I held with him.

Last but not least, I want to thank very warmly Mrs Jadwiga Krasnowska, the typist whose patient work with the manuscript of both the versions of the book was an invaluable help for me.

November 1977

RYSZARD WÓJCICKI

PREFACE TO THE POLISH EDITION

Development of logic greatly contributed to establishing of a particular approach to mathematical theories applied in methodological studies. Under this approach mathematical theories are treated as idealized counterparts of certain branches of mathematics, describing certain idealized objects called models or realizations of theories. From the formal standpoint empirical theories or at least some of them do not differ from mathematical ones. Already for this single reason both the way of handling an object of inquiry and the means used in mathematical logic can be conveyed to all theories in general, and thus in particular to empirical disciplines. The methodology based on such an approach is called logical theory of science.

As a matter of fact, although I would prefer to understand the name "formal methodology" in a somewhat broader sense so as to make it include all researches into the structure and dynamics of theories carried out with the help of any mathematical means, i.e. not necessarily reduced to those employed in mathematical logic, the considerations to be found in this book do not go far beyond the scope of logical theory of science. Making the decision as to the title of the book I was quite conscious that it may suggest too much, but on the other hand I am convinced that the discussion to be found here concerns the problems fundamental for any formal considerations which deal with cognitive processes. Since an introductory discussion of the problems considered in the monograph can be found in Chapter 1, now I would like to devote some space to certain special questions closely related to those investigated in the book.

One of the questions often raised in connection with objective of logical theory of science is to what extent the theory can go beyond mathematical logic. It is sometimes suggested that logical theory of science is nothing else but reinterpretation of the results of metamathematics so as to make them applicable to methodological considerations in a wide sense. Were the problem reduced just to the interpretation of results of mathematical logic in this sense, still it would not have to mean that such an operation is trivial or redundant. As a matter of fact, however, it is peculiar to empirical sciences that not only the results of mathematical logic often gain a new essence when applied to empirical sciences but some specific problems having no counterparts in metamathematics arise there too. For instance, the whole measurement theory, although exceeds the frameworks of metamathematics, obviously belongs to logical theory of science due to the fact that it exploits the same formal tools as mathematical logic does. Another example of such a specific problem characteristic of methodology of empirical science is analysis of the notion of regularity. And again, mathematical logic provides formal apparatus by means of which the analysis can be performed, although the notion of regularity has hardly anything in common with metamathematics. A very significant example of the methodological problem having no counterpart among problems of fundamentals of mathematics is the one of approximate character of empirical inquiry. On passing to general methodology, the notion of truth employed in metamathematics must be replaced by the notion of approximate truth.

When discussing relations between formal methodology and mathematical logic one should bear in mind one more interdependence of considerable significance for the place occupied by formal methodology among the disciplines whose main concerns are science and cognitive processes. I mean here interrelations between methodology and epistemology. It is quite obvious that formal methodology can be treated as a branch of epistemology. Problems covered by the former are nothing else but a fragment of the whole set of problems dealt with by the latter. And yet, due to the formal approach a crack between the epistemology and methodology is becoming to be clearly visible. The point is that each area of inquiry, each scientific discipline, is characterized not only by the problems which are being investigated in its framework, but also by the set of methods employed by the discipline. It seems to be quite justified to speak about a germ of certain emancipation of methodology from philosophy similar to that attained by sociology, psychology, and logic some time ago.

I would dislike this remark to be treated as an attempt to separate formal methodology from philosophy. Existence of certain individual features by no means excludes existence of close correlation. It is nowadays hardly imaginable that a contemporary philosopher being interested in epistemological problems ignores the results of psychology, logic or just logical theory of science. But, on the other hand, one can hardly think of a methodologist who would take into account no philosophical assumptions whatsoever in his work. Perhaps that could happen if methodology of science were built up as a sum of inductive generalizations of the particular cases examined. But, similarly to metamathematics and in fact to any science making use of mathematical techniques, formal methodology tends to construct ideal models for those objects and phenomena (theories and changes they undergo) which are subject to its researches. Collected data as well as inductive generalizations based on them certainly may, and as a matter of fact must constitute a point of departure for constructing formal models. Theoretical generalizations, however are always characterized by some autonomy with respect to the evidence on which they are based. Constructing a formal model we usually consciously neglect certain features of the phenomenon under modelling in spite of their repetitive character treating them as secondary, and bring others into prominence, although no inductive scheme may happen to be capable of detecting them. Various and often purely subjective factors may condition the choice of this and not some other way of searching for a sound theoretic generalization. It is philosophical conceptions which constitute the factor playing an inspiring, guiding and often decisive role in the search for suitable research lines in methodological investigations.

On the other hand, verification of formal models consisting in checking their applicability to the description of actual cognitive procedures may be a departure point for verification of the philosophical ideas on which the models were based; most often philosophical conceptions are formulated in such a general and loose way that estimation of their soundness cannot be directly performed. That is how methodology can pay its debt to philosophy. Let us illustrate the above remarks with some examples which, moreover, will, to some extent of course, provide fragmentary characteristic of philosophical assumptions on which the considerations to be found in this book have been based.

Although perhaps it ought to be regretted, the time of paradoxical ontological conceptions seems to belong already to the past. Philosophy entered minimalistic, common-sensical, and thus unfortunately rather dull, stage. At present one can hardly find a methodologist who would not wholly accept the thesis on objective character of cognition. Clearly, this thesis is one of the fundamental assumptions of the present work. However, the point is that the persuasion about the objective character of cognition can mean different things with different authors. I shall mention here two largely opposing and in a sense radical interpretations given to it – both of them having very little in common with the assumption accepted in the book.

First of the two approaches reduces to the opinion according to which the world accessible to cognition is a world of the objects which can be seen, touched, weighed, measured. This approach equate reality with observable reality. The typical representative of this approach is a philosophical trend called logical empiricism or sometimes neopositivism. I doubt whether one can, basing on the texts, give an unquestionable argument to the effect that the representatives of logical empiricism deny existence of nonobservational, i.e. theoretical objects. The problem of existence of theoretical objects is not, as a rule, undertaken by them on ontological ground. However, reduction of the entire knowledge to the observational level postulated by logical empiricism is in fact equivalent to acceptance of the assumption that what is not directly observable is not accessible to any inquiry. Consequently the question of existence or nonexistence of theoretic objects is treated as of purely speculative character. Unless the line separating observational from theoretical terms is defined in rigorous terms, the account of science based on that dychotomy – often called the partial interpretation view – can hardly be exposed to any evaluation. Still, I am afraid, under any acceptable specification the partial interpretation view involves drastic oversimplications.

Let me try to explain in an outline what underlies my belief stated above. There seems to be not too much hope to find the criteria which allow us to draw the borderline between observational and theoretical terms in a clear-cut way. Yet, if in some way we manage to do so, we obtain a white-black picture with the terms considered as observable on one side of it and theoretical ones on the other. In fact, however, as it was pointed out by numerous critics of the partial interpretation view, although there is no reason to question existence of purely observational terms and there is no reason to deny existence of purely theoretical terms, an overwhelming majority of terms involve in their meaning both theoretical and observational components. Thus if indeed theoretical and observational attributes inhere and intertwine with each other in the connotation of nearly all terms used in science, almost each decision of the form "this term is theoretic" or "this term is observable" is erroneous. The errors being cumulated, the entire picture of science becomes distorted.

How completely different the viewpoints of the philosopher subscribing to the partial interpretation view and of the physicist are may be well illustrated by the following example. Ask both of them whether the sentence "lifetime of the meson π° is approximately 10^{-16} sec", is an empirical fact. The physicist's answer is very likely to be: – Yes, the fact is perfectly well established empirically: traces of the meson π° in photographic emulsion are observable in microscope and their length is measurable. The follower of logical empiricism on the other hand, will notice that neither the meson π° nor the time 10^{-16} sec belongs to observables and consequently will consider the theorem on the meson to be a theoretical one and requiring reduction to the observational level.

What exactly the reduction is to be, how deep we should descent to reach the observational bottom, since, as all seems to indicate, photographic emulsion and microscope are theoretical terms – nobody knows. Successive definitions of empirical meaningfulness given by Carnap turned out to be successive failures, and in the whole very rich literature on philosophy which took the requirement of strictness and verifiability as its basic watchward, one is unable to find a single example which would illustrate in a satisfactory way how theorems of physics may be established on the basis of the facts devoid of any theoretic tinge. Everything seems to indicate that the partial interpretation view has been a blind alley in the development of philosophy of science. It should only be regretted that by rather perfunctory associations failures of logical empiricism are thought of as failures of logical theory of science. The considerations concerning the latter can be based on various philosophical assumptions: there is no inevitable *junctim* between logical empiricism and logical theory of science.

The critical remarks presented above may become a source of a certain misunderstanding. While criticizing logical empiricism I by no means intend to reject the thesis about empirical origin of knowledge we have. Asking, however, how the knowledge is grounded on experience I am inclined to trust the physicist's intuitions, who clearly plays here the role of the representative of empirical sciences, rather than those of the philosopher's.

I think, in particular, that the methodologist should at least consider the possibility of accepting the idea of an empirical fact as it is understood by scientists, as a starting point for his research. I feel quite conscious of numerous objections that can be set forth against my suggestion. One of them can probably be as follows. Empirical facts serve for verification of a theory, while representatives of sciences count as empirical facts only the statements obtained on the ground of the theory. Of what value can verification of the theory be, if performed by means of results obtained with devices designed on the basis of the theory under verification? Don't we face a vicious circle here? Let me present my attitude assumed toward this argument which, although roughly outlined, is, I hope, quite clear.

In general the fact that a device was constructed on the basis of laws of the theory does not guarantee that the results obtained with the help of it will be consistent with predictions of the theory.

It may happen that they contradict the theory. What then? The answer is that reliability of measurement being not questioned, we shall modify the theory which can, although not has to, yield the necessity to verify theoretical grounds of the performed measurement and the necessity to reinterpret the obtained results. The modifications can be thus pretty for going. They must result in consistency of the theory and experience. Observe that although the theory and experience are so closely interconnected, it does not mean that we are unable to trace how experimental data affect the construction of the theory.

Now let us pass over to the other approach connected with the thesis about the objective origin of our knowledge. Due to my empirical credo, I fiercely object against the ontologies which without any restraints fill the world with the objects inaccessible in principle to any empirical cognition. Specially inspiring role in this respect seems to be played by idealizing notions and theorems, so typical for science. The reasoning is as follows. Guy-Lussac's law does not describe real gases, well – never mind, it refers to ideal gas. No real body behaves exactly as it is predicated by laws of theoretical mechanics – never mind, theoretical mechanics describes particles and ideally rigid bodies. The laws of political economy do not work in its pure form in any of the existing economical systems – never mind, they describe some ideal systems, and so on, and so forth.

Nobody seems to deny that apart from ideal objects and ideal systems there are real empirical phenomena, but methodological analysis of relations between theories and experience is broken half-way, usually by some pseudosolution of the problem. One of such pseudosolutions is a fairly widespread conception of counterfactual. Thus it is sometimes maintained that indeed Guy-Lussac's law is not true with respect to real gases but it is counterfactually true with respect to them, i.e. it would be true if the gases satisfied certain conditions which they in fact do not satisfy. In this case the conditions are: particles of the gas are dimensionless points not interacting among themselves. Unfortunately, I am afraid that the notion of a counterfactually true sentence is misleading. If indeed we are able to say something meaningful about how a given gas would behave if it satisfied the two conditions stated above, it is only on the ground of such laws as that of Guy-Lussac's. In other words, the only reasonable interpretation of the counterfactual seems to be the one under which Guy-Lussac's law is counterfactually true because it is a theorem of physics. No matter how troublesome it could be, theorems on ideal gases, dimensionless particles, ideally rigid bodies or any abstract objects which never occur in their "pure" form are in fact theorems on real gases, real physical bodies, in general on real objects existing in the real, accessible to our cognition world.

I hope that the remarks above provide an effective defence against a possible charge that I underestimate the role of the philosopher. It is just the opposite. But I think that philosophical assumptions demarcate only the most general line along which the methodological researches should be carried out, and it is only the specification of these assumptions reached as a result of constructing formally well-defined models which leads to verifiable, accessible to estimation constructions. Criteria for deciding whether a proposed model should be accepted or rejected are always the following: the degree of accuracy with which the constructions reflect properties of real objects, and the degree of susceptibility to further improvements.

Finally I wish to add some words on the character of the publication. In spite of a considerable progress which influenced many of its branches, methodology which operates with formal tools is still a conglomerate of partial results obtained on the ground of not always convergent intuitions, research lines and ways of thinking. In this situation no attempt of presenting methodological problems in a wider sense can be of standard and independent of the author's personal opinions character. In this sense, the present work is obviously marked with subjectivity. This does not mean, of course, that the book was written quite separately from publications by other authors. It would not at all be possible. After all, apart from decisions based on certain controversial assumptions, logical theory of science contains also the results which are well-grounded and generally accepted. All studies conceived on a wider scale must refer to these results. And thus for instance the point of departure for the analysis of an approximate notion of truth presented in the book is the known Tarski's conception, the discussion on the notion of a quantity and considerations concerning the theory of measurement are based on the results which can be found in the already vast literature on the problems in question, studies on the notion of regularity refer to certain definitions known in cybernetics.

Quite a different matter is the problem of inspiration which I gained from works of various authors or finally convergence of some of the results included in this book with those to be found in publications of other authors. By suitable comments or references I tried to call the reader's attention to the publications in which one can find discussion on the problems related to those discussed here. Of course, the bibliographical remarks I give do not claim to be exhaustive.

For two main reasons I aimed at as complete exposition of the problems discussed as possible. Since it is still the fact that in the area of methodology divergent conceptions are being developed, it becomes a matter of considerable importance to present intuitive bases of the discussion to be carried out. In case the author confines himself to purely formal considerations, it becomes the reader's task to find for them substantially proper interpretation, and then misunderstandings are very likely to occur.

Quite independently of the attempt to provide the reader with the appropriate interpretational key, I tried to make the book intelligible not only for philosophers of science. Methodological reflection accompanies any research work. In spite of their general and abstract nature, the problems in the book should be of interest for people working in "object-science" not just "meta-science". Propagation of the notions and ideas of formal methodology of science is certainly the aim worth striving for.

I. INTRODUCTION

1.1. MATHEMATICAL NOTATION

The logical and mathematical notation applied in this book will be standard, and the reader who is familiar with rudiments of set-theory, mathematical logic and elementary mathematics should find it intelligible without any explications. Still in this section we shall provide a short survey of symbols which will be in the most often use. The survey is by no means thought to be an exhaustive one. We merely wish to make a few preliminary remarks on the conventions to be followed.

First, we shall use the common logical connectives in their truth-functional sense: "if ... then" is the connective of material implication, "not" of negation and so on. We shall avoid symbolic notation. The logical symbols $\forall$, $\exists$, $\wedge$, $\vee$, $\neg$, $\rightarrow$, $\leftrightarrow$ will be used only in some special contexts. They stand for: *for all, for some, and, or, not, if ... then, if and only if,* respectively.

The symbols: $\in$, $\notin$, $\times$, $\cap$, $\cup$, $\bigcup$, $\bigcap$, $\varnothing$, $\{a_1, \ldots, a_n\}$ will have their usual set-theoretical meaning, thus, for instance, given two sets A and B, $A \times B$ is the *Cartesian product* of these sets, $A \cup B$ is their *union*. If K is a set of sets, $\bigcup K$ is the *union* and $\bigcap K$ is the *intersection* of its elements. $\varnothing$ denotes the *empty set*. $\{a_1, \ldots, a_n\}$ is the set whose elements are $a_1, \ldots, a_n$. Only the symbol of inclusion will be applied in the way which is not commonly accepted. Namely, $\subset$ will denote the *proper inclusion* of two sets (so if $A \subset B$, then $A \neq B$), the *inclusion* which need not be proper will be denoted by $\subseteq$.

If a function f maps A into B, we shall write $f: A \rightarrow B$, and sometimes also $a \rightarrow f(a)$, where $a \in A$. The set A will be called the *domain* and the set B the *range* of f. If A is of the form $A_1 \times A_2 \times \ldots \times A_k$, the set A_i will be said to be the *i-th domain* of f. Similarly in the case of a k-ary relation $R \subseteq A_1 \times A_2 \times \ldots A_k$ we shall call A_1 its first domain, A_2 its second domain and so on.

The symbol *Re* will be applied to denote the set of all real numbers. We shall occasionally apply Re^+ and Re^- as symbols for the set of all positive real numbers and that of all negative real numbers, respectively. 0 will be treated as both positive and negative real number.

The symbol $[Re]$ will stand for the set of all non-empty and closed intervals of real numbers. The elements of $[Re]$ are then intervals which possess one of the following forms: $(\infty, x]$, $[x, \infty)$, (∞, ∞), $[x, y]$, where $x, y \in Re$. We shall assume that $[x, y] = [y, x]$, i.e. the two symbols denote the same interval. Clearly if $x \leqslant y$ then $[x, y] = [y, x] = \{z \in Re: x \leqslant z \leqslant y\}$. Incidentally, the symbol $\{z \in Re: x \leqslant z \leqslant y\}$, or in an abbreviated notation $\{z: x \leqslant z \leqslant y\}$, illustrates an application of the definitional schema $\{x: \Phi(x)\}$ (the set of all those x which satisfy the condition $\Phi(x)$) which will be often used.

The symbol $\mathcal{N}$ will stand for the set of all natural numbers.

Let Λ be a set of integers of the form $\{i: i_1 \leqslant i \leqslant i_2\}$, and let X be an arbitrary set of elements. The function

$$\sigma: \Lambda \to X$$

will be said to be a *sequence* of elements of X. If Λ is finite the sequence σ will be called *finite* or else *infinite*. Finite sequence will be also referred to as *ordered sets* and often presented in the form

$$(a_1, \ldots, a_k).$$

Infinite sequences of the form

$$a_0, a_1, a_2, \ldots, a_i, \ldots$$

i.e., sequences $\sigma: \mathcal{N} \to X$, such that $\sigma(i) = a_i$, will be denoted, as it is customary, as $\{a_n\}$. The symbol $\{a_n\}$ abbreviates the symbol $\{a_n\}_{n \in \mathcal{N}}$, and the latter is an instance of the symbol

(1) $$\{a_\mu\}_{\mu \in M}$$

where M is an arbitrary set and all objects of the form a_μ are elements of a set X. As known, (1) denotes the function $\mu \to a_\mu$, i.e., the function which assigns to each element μ of M the element a_μ of X.

The reader should perhaps be warned that (1) and

(2) $\{a_\mu \colon \mu \in M\}$

do not stand for the same thing. The latter symbol denotes the range of the function (1), i.e. (2) is the set of all objects of the form a_μ where μ is in M. The sets given in the form (2) are often called *indexed sets*.

If $K = \{A_\mu \colon \mu \in M\}$ is a family of sets, we shall often write $\bigcup A_\mu$ instead of $\bigcup K$, and similarly $\bigcap A_\mu$ instead of $\bigcap K$. $\bigcup_{\mu \in M} A_\mu$ and $\bigcap_{\mu \in M} A_\mu$ are still other notational variants which sometimes will be applied.

Similar conventions are to be followed in the case of all operations defined on infinite sets. If for instance $\{a_n\}$ is a sequence of real numbers, the sum of the elements of that sequence will be denoted either as $\sum_{n \in \mathcal{N}} a_n$, or more often, as $\sum_{n=1}^{\infty} a_n$. Occasionally we may simply write $\sum a_n$.

1.2. NUMERICAL SYSTEMS

A notion which will be especially often used in our further considerations is that of a *structure*, i.e. an ordered set of the form

(1) $\mathfrak{A} = (A_1, \ldots, A_m, R_1, \ldots, R_k)$

where $A_1, \ldots, A_m$ are non-empty sets and $R_1, \ldots, R_k$ are relations defined either directly on some of these sets or on some set--theoretical constructs formed from them. We shall say that X is a *set-theoretical construct* formed from $A_1, \ldots, A_m$ if and only if there exists a finite sequence

$$X_1, X_2, \ldots, X_m$$

such that $X_m = X$ and each X_i in the sequence either coincides with some of the sets $A_1, \ldots, A_n$ or results from some of the elements preceding it in the sequence under any of the standard set-theoretical operations such as forming union of sets, Cartesian product, power set (i.e. the set of all subsets of a given set), etc.

A rigorous definition of a set-theoretical construct and thus also a structure is easily available, but somewhat incomplete definitions provided above will suit our purposes satisfactorily well.

A structure of the form (1) will be called *m-sorted*. The sets A_i will be called *domains* of $\mathfrak{A}$ and the set $\{R_1, \ldots, R_k\}$ of the relations involved by $\mathfrak{A}$ will be referred to as *characteristic* of $\mathfrak{A}$. If $R_1, \ldots, R_n$ are relations defined on $A_1, \ldots, A_m$, (e.g. $R_1 \subseteq A_1 \times A_1$, $R_2 \subseteq A_2$, $R_3 \subseteq A_1 \times A_7 \times A_8, \ldots$), then (1) will be said to be an *m-sorted relational structure*. A one-sorted relational structure will be referred to as a *relational structure*.

Structures of the form

$$(A, f_1, \ldots, f_n),$$

where f_i are operations on A will be referred to as *abstract algebras*. (By an *n-ary operation* on A we mean a function $f\colon A^n \to A$). Clearly, functions, and thus operations in particular, can be treated as relations of a certain special sort. The terminology we presented does not depart from that being in common use.

Observe that often it is enough to define the characteristic of $\mathfrak{A}$ in order to define some or even all of the domains of $\mathfrak{A}$. For instance if $\mathfrak{A}$ is an algebra, the domain A of $\mathfrak{A}$ is the domain of the operations f_i. The definition of A can be then regarded to be a part of the definition of any of f_i' s. If the definition of a domain A_i of $\mathfrak{A}$ can be reconstructed from the definition of the characteristic of $\mathfrak{A}$, we can drop out A_i from the constitutents of $\mathfrak{A}$. Thus for instance the algebra $(A, f_1, \ldots, f_n)$ may be presented in the form $(f_1, \ldots, f_n)$, the relational structure $(A, R_1, \ldots$ $\ldots, R_n)$ in the form $(R_1, \ldots, R_n)$. We shall occasionally make use of that possibility.

The term *system* will be applied as a synonym of "structure". It will also serve (especially in this and the next chapters) as an abbreviation for "a numerical system", i.e. a structure of a certain special kind we are going to define now.

Consider any set of objects A, an interval u of real numbers and functions $F_1, \ldots, F_n$ of the form

$$F_i : u \times A^k \to Re,$$

where $k = k(i)$. Then

(2) $$\mathfrak{A} = (A, u, Re, F_1, \ldots, F_n)$$

is a three-sorted structure of a relatively simple kind. Just the structures of the sort defined will be referred to as *numerical systems*, or *systems* for short.

In what follows we shall often assume that numerical systems we deal with are empirical systems in the sense that the functions F_i are certain quantities whose values can be measured by means of suitable empirical procedures. Note that this interpretation does not presuppose anything about the nature of elements of $\mathfrak{A}$. A may comprise physical bodies, points of physical space, sets of empirical bodies or even certain structures.

When $\mathfrak{A}$ is assumed to be an empirical structure, the interval u will be interpreted as a period of time: the period of existence of $\mathfrak{A}$ or just the period in which $\mathfrak{A}$ is examined. It will be referred to as *duration* of $\mathfrak{A}$. Consequently, elements of u will be referred to as *instants*. Thus the formula

(3) $$F_i(t, a_1, \ldots, a_k) = x$$

reads: *the quantity F_i measured on $a_1, \ldots, a_k$ at time t has the value x.*

In the sequel we shall denote the duration u of $\mathfrak{A}$ as $T(\mathfrak{A})$ and present $\mathfrak{A}$ in the form

(4) $$\mathfrak{A} = (A, F_1, \ldots, F_n)$$

rather than in (2).

By an *element* of $\mathfrak{A}$ we shall understand an element of A. If A is finite, the system $\mathfrak{A}$ will be said to be *finite* or else *infinite*.

The functions F_i will be called *system variables* of $\mathfrak{A}$. They should not be confused with language variables e.g. such as those which in formula (3) appear as: *time argument* t, *object arguments* $a_1, \ldots, a_n$ and *real value* x.

Variables (functions) which do not belong to the characteristic of a system will be called *parameters*, especially if they affect the values that the system variables may take.

The variables a numerical system involves need not have the whole set Re as their ranges, in particular they may happen to be characteristic functions of some relations. (The function f_R is said to be the *characteristic function* of the relation $R \subseteq A_1 \times \ldots \times A_k$ if and only if $f_R : A_1 \times \ldots \times A_k \to \{0, 1\}$, and $f_R(a_1, \ldots, a_k) = 1$ if and only if $R(a_1, \ldots, a_k)$). Clearly a system $(A, f_{R_1}, \ldots, f_{R_n})$ whose characteristic consists of characteristic functions of the relations $R_1, \ldots, R_n$ can be treated as identical with the relational structure $(A, R_1, \ldots, R_n)$. We shall occasionally refer to systems of the form described as *pseudo-quantitative*.

Observe also that the dependence on time of the system variables can be apparent. It may happen that for all system variables F_i of $\mathfrak{A}$, for all t, t' in $T(\mathfrak{A})$, and for all $a_1, \ldots, a_k$ in A,

$$F_i(t, a_1, \ldots, a_k) = F_i(t', a_1, \ldots, a_k).$$

In that case the system $\mathfrak{A}$ will be said to be *stationary* or else *dynamic*.

Each structure of the form (4) where each F_i is a real-valued function of the form

$$F_i(a_1, \ldots, a_k) = x,$$

(i.e. is a function which depends only on object arguments and does not depend on time) can be treated as a stationary system. Observe however that duration of such a system is not uniquely defined.

Given a system $\mathfrak{A}$ of the form (2) (or equivalently (4)), the sequence

$$(k(1), \ldots, k(n)),$$

i.e., the sequence of the numbers of object arguments of the functions $F_1, \ldots, F_n$, will be called the *similarity type* of $\mathfrak{A}$.

Two systems of the same similarity type will be called *similar*. Thus $\mathfrak{A}$ and a system

(5) $$\mathfrak{B} = (B, G_1, \dots, G_m)$$

are similar if and only if $n = m$ and the corresponding variables F_i and G_i take exactly the same number of arguments.

Given two similar systems of the form (4) and (5) respectively we shall say that $\mathfrak{B}$ is a *subsystem* of $\mathfrak{A}$ if and only if $B \subseteq A$, $T(\mathfrak{A}) \subseteq T(\mathfrak{B})$, and each parameter G_i is the restriction of the corresponding parameter F_i simultaneously to B and $T(\mathfrak{B})$, i.e., for all t in $T(\mathfrak{B})$ and for all $a_1, \dots, a_k$ in B

$$F_i(t, a_1, \dots, a_k) = G_i(t, a_1, \dots, a_k),$$

$i = 1, \dots, n$.

If $T(\mathfrak{A}) = T(\mathfrak{B})$, the subsystem $\mathfrak{B}$ will be denoted as $\mathfrak{A}/B$. If $A = B$, the subsystem $\mathfrak{B}$ will be denoted as $\mathfrak{A}/T(\mathfrak{B})$ and will be referred to as a *time-restriction* of $\mathfrak{A}$ to $T(\mathfrak{B})$.

Two similar systems $\mathfrak{A}$, $\mathfrak{B}$ which differ only as to the variables they involve, i.e. such that $A = B$ and $T(\mathfrak{A}) = T(\mathfrak{B})$, will be referred to as *variants* of each other.

Any system $\mathfrak{B}$ which results from $\mathfrak{A}$ by removing some of the variables the latter involves will be called a *reduct* of $\mathfrak{A}$; at the same time $\mathfrak{A}$ will be called *expansion* of $\mathfrak{B}$.

The following two concepts will be of special importance.

DEFINITION 1.2.1.

A. *A system $\mathfrak{B}$ is said to be a* homomorphic image *of $\mathfrak{A}$ under f if and only if $f: A \to B$, and there exists a real number r such that the following conditions hold true:*

(i) *$f(A) = B$, i.e. f maps A onto B*

(ii) *for each real number t, t is in $T(\mathfrak{A})$ if and only if $t+r$ is in $T(\mathfrak{B})$*

(iii) *for each real number x, for each t in $T(\mathfrak{A})$ and for each sequence $a_1, \dots, a_k$ of elements of $T(\mathfrak{A})$,*

$$F_i(t, a_1, \dots, a_k) = G_i(t+r, fa_1, \dots, fa_k),$$

where $i = 1, \dots, n$.

B. *If moreover f is one-to-one,* $\mathfrak{B}$ *is said to be an* isomorphic image *of* $\mathfrak{A}$ *under* f, *in symbols* $\mathfrak{A} =_f \mathfrak{B}$.

As it is readily seen, when f establishes isomorphism of $\mathfrak{A}$ to $\mathfrak{B}$, the inverse function f^{-1} establishes isomorphism of $\mathfrak{B}$ to $\mathfrak{A}$.

NOTE. Under the interpretation we ascribe to the notion of a system variable and similarly to that of a parameter, any quantity which is defined on a set of objects, depends on time and takes real numbers as its values may, when referred to in a suitable context, serve as an example of either a system variable or a parameter.

There are several reasons why quantities involved in empirical theories need not, unless redefined in an appropriate manner, fall under the category of quantities mentioned above.

First, values of some of empirical quantities are more sophisticated objects than just real numbers, they can be e.g. vectors. Such quantities however can be, as a rule, reduced to real-valued ones. Each vector in the physical space, for instance, can be represented uniquely by a triple of scalars.

Some of the quantities are functions of intervals of time, say annual salary or average temperature during a season. But again they are usually definable in terms of quantities whose values are defined for each single instant of time. Annual salary of an employee is the sum of all payments he received as a salary at any time during a year. A definition of an average temperature is slightly more complicated, it is clear however that the average temperature is a function of the temperature at a given time.

The value which a quantity takes may depend on a selection of unit of measure and frame of reference. The ordinary practice here is to add an indication of the unit of measure, e.g. cm, sec, or $, to the number expression for the value of the quantity and to mention the frame of reference in a separate comment. Both the specification of the unit and that of the frame of reference may be treated as a part of definition of the quantity. Thus the symbol v need not stand for, say, velocity but for velocity measured in centimeters with respect to a given frame of reference *r*. Again the procedure we have described can be applied in order to redefine a quantity so as to obtain the one which can serve as a system variable.

All that has been said above to supplement the relevant parts of Chapter I does not go beyond the standard comment on the notion of an empirical quantity which can be found in a number of writings (cf. e.g. J. von Neuman and O. Morgenstein, 1953, p. 22; R. Carnap, 1958, Section 41, 1966; M. Bunge, 1967, p. 50).

We shall conclude with certain notational remarks. It is customary in physics, and generally in empirical sciences, to avoid the notation which exhibits explicitly all arguments on which a given variable depends. Thus for instance, the physicist writes *Second Newton's Law* as follows

(1) $$F = m \frac{d}{dt} s$$

which, in its non-abbreviated form, amounts to something like: *for all particles p in P and for all instants of time t' in T*:

(2) $$F(t', p) = m(t', p) \frac{d^2}{dt^2} s(t, p)_{t=t'}.$$

Compare e.g. the way in which this law has been presented in the axiomatic formulation of Particle Mechanics given by I. C. McKinsey, A. C. Sugar and P. Suppes (1953), see also P. Suppes (1957). The symbols P and T which appear in the quantifying expressions preceding (2) refer to the set of elements and to duration of the systems to which the law can be applied and acquire a definite interpretation provided that such a system has been pointed out.

1.3. ELEMENTARY SYSTEMS

In our further considerations an especially important role will be played by systems called elementary. We shall define a system to be *elementary* if and only if the set of its elements is a *unit set*, i.e. a set of the form $\{a\}$.

Throughout the book elementary systems will be denoted by small gothic letters $\mathfrak{a}, \mathfrak{b}, \mathfrak{c}, \ldots$ and thus the system

(1) $$(\{a\}, F_1, \ldots, F_n)$$

or equivalently

(2) $$(a, F_1, \ldots, F_n)$$

will be denoted as $\mathfrak{a}$ rather than $\mathfrak{A}$.

Since there exists only one element a with respect to which variables of the system $\mathfrak{a}$ are defined, they are always functions of the form

(3) $$F_i(t, a)$$

It is obvious, however, that when a fixed system $\mathfrak{a}$ is considered, a becomes an apparent argument; it does not vary and hence does not influence values of parameters of the system. We may then consider each elementary system as a system of the form

$$(4) \qquad \mathfrak{a} = (F_1, \ldots, F_n),$$

the variables F_i being functions of the form

$$(5) \qquad F_i(t) = x.$$

In what follows the notation given in (4) and (5) will be most frequently used. It will always be treated as abbreviated, however. We shall assume that whenever a system has been presented in the form (4), there exists an object a such that the variables F_i are defined with respect to this particular object a, and thus they are, although implicitly, functions of the form (3).

The motivation for this decision is simple. Throughout this book we shall be especially interested in systems which represent certain empirical phenomena (more about that in Section 1.6). Now, suppose that we compare behaviour of two elementary systems which involve "the same" variables, say, we are studying changes in number of two populations p and q of animals, and the only variables in which we are interested are the number of males N_m and that of females N_f in a given population. We deal then with two elementary systems:

$$\mathfrak{p} = (p, N_m, N_f),$$
$$\mathfrak{q} = (q, N_m, N_f).$$

It may happen that for each t in the time interval within which the systems are studied the following identities hold true:

$$N_m(t, p) = N_m(t, q),$$
$$N_f(t, p) = N_f(t, q).$$

It does not mean, however, that after having simplified the notation by omitting the symbols p and q we may treat the systems

$$(6) \qquad \mathfrak{p} = (N_m, N_f),$$
$$(7) \qquad \mathfrak{q} = (N_m, N_f)$$

as identical. The way in which the systems $\mathfrak{p}$ and $\mathfrak{q}$ have been presented in (6) and (7) respectively is, in fact, misleading. It conceals the fact that the functions N_m and N_f involved in (6) and those involved in (7) are defined with respect to different populations.

One may suggest to write N_m^p, N_f^p, and N_m^q, N_f^q, respectively, in order to distinguish variables of one system from corresponding parameters of the other. We shall often follow this reasonable suggestion. But again we have:

$$N_m^p(t) = N_m^q(t)$$

and

$$N_f^p(t) = N_f^q(t)$$

which clearly proves that $N_m^p = N_m^q$, and $N_f^p = N_f^q$. If we regard, however, p and q as implicit parameters on which the systems (6), (7), or equivalently

$$\mathfrak{p} = (N_m^p, N_f^p),$$
$$\mathfrak{q} = (N_m^q, N_f^q),$$

depend, we shall treat those systems as isomorphic rather than identical.

A numerical system which is not elementary will be called *generic*. All numerical systems are thus divided into two categories: elementary and generic.

The importance of the notion of an elementary system is due to the fact that each finite system can be uniquely represented as an elementary one. Although this assertion can easily be stated in a complete and precise manner, let us confine ourselves to a short example which should make it clear what we have in mind.

Let

$$\mathfrak{A} = (\{a_1, a_2, a_3\}, F, G)$$

be a three element system and let F depend on one and G on two object arguments. Put

$$F_i(t) = F(t, a_i),$$
$$G_{ij}(t) = G(t, a_i, a_j),$$

where i and similarly j may take any of the values $1, 2, 3$, and t runs over $T(\mathfrak{A})$.

As we easily agree, the system $\mathfrak{A}$ and the elementary system

(8) $$\mathfrak{a}_{\mathfrak{A}} = (F_1, F_2, F_3, G_{11}, G_{12}, \ldots, G_{33})$$

are, in a sense, equivalent. Variables of the latter have been defined in terms of variables of the former (more exactly in terms of variables and elements of that system), and, as it is immediately seen, values of the variables of $\mathfrak{A}$ are definable in terms of values of those of the system $\mathfrak{a}_{\mathfrak{A}}$ as well.

The notation in (8) is an abbreviated one, and one may ask what the implicit object $a_{\mathfrak{A}}$ with respect to which the variables of the system $\mathfrak{a}_{\mathfrak{A}}$ are defined, is like. Clearly

$$a_{\mathfrak{A}} = \{a_1, a_2, a_3\},$$

i.e. it is the set $\{a_1, a_2, a_3\}$ considered as a whole.

It is rather obvious that the method by means of which the system $\mathfrak{a}_A$ was constructed as a counterpart of $\mathfrak{A}$ can be applied in the case of any finite generic system $\mathfrak{B}$. However, if $\mathfrak{B}$ was infinite, say, it was a system whose elements are space points of a given region of a physical space, it could not be transformed onto an elementary system in the way described, since the resulting "system" would involve an infinite number of variables.

1.4. STATES AND HISTORIES

We shall use the symbol $\mathfrak{A}(t)$ to denote a state of the system $\mathfrak{A}$ at time t. A definition of $\mathfrak{A}(t)$ takes especially simple form when $\mathfrak{A}$ is an elementary system. Let us consider then an elementary system

$$\mathfrak{a} = (F_1, \ldots, F_n)$$

first. The *state* of $\mathfrak{a}$ at time t is the sequence:

(1) $$\mathfrak{a}(t) = (x_1, \ldots, x_n),$$

where $x_i = F_i(t)$, i.e., it is the sequence of the values which state variables take at that time.

In the general case the definition must take a more complicated form, for if $\mathfrak{A}$ is a generic structure, values of the variables of that system depend not only on time but also on object arguments.

To make the definition as simple as possible, we shall adopt the following approach. Given a structure

$$\mathfrak{A} = (A, F_1, \ldots, F_n)$$

and given an arbitrary t in $T(\mathfrak{A})$, we define first a new parameter F_i^t by postulating:

$$F_i^t(a_1, \ldots, a_k) = F_i(t, a_1, \ldots, a_k),$$

and next we put:

$$\mathfrak{A}(t) = (F_1^t, \ldots, F_n^t). \tag{2}$$

Observe that since t is fixed, the parameters F_i^t do not depend on time.

Verify that the following holds true:

ASSERTION 1.4.1. $\mathfrak{A}(t) = \mathfrak{A}(t')$ *if and only if for all system variables* F_i *and for all* $a_1, \ldots, a_k$ *in* A,

$$F_i(t, a_1, \ldots, a_k) = F_i(t', a_1, \ldots, a_k).$$

In view of the fact stated above the state of the system at time t is uniquely determined by the values which the system variables take at that time for all possible object arguments in A. This may regarded as an argument in favour of the adequacy of the definition.

One may easily verify that if $\mathfrak{A}$ is an elementary structure, the right-hand side of (2) becomes identical with (1), and thus, as one should expect, the definition of a state of an elementary system turns out to be a special case of the general one.

The symbol $St(\mathfrak{A})$ will denote the set of all states of $\mathfrak{A}$. We put then:

$$St(\mathfrak{A}) = \{\mathfrak{A}(t) : t \in T(\mathfrak{A})\}.$$

A state of a variant of $\mathfrak{A}$ will be called a *possible state* of $\mathfrak{A}$. The set of all possible states of $\mathfrak{A}$ will be denoted by $St^*(\mathfrak{A})$. In accordance with the definition stated above

$$St^*(\mathfrak{A}) = \bigcup \{St(\mathfrak{B}): \mathfrak{B} \text{ is a variant of } \mathfrak{A}\}.$$

The function

$$\{\mathfrak{A}(t)\}_{t \in T(\mathfrak{A})}$$

will be referred to as a *history* of the system $\mathfrak{A}$. In fact we then identify the system and its history. If systems are isomorphic we shall say that their histories coincide up to the translation in time with the help of which the isomorphism has been established.

Let us express this definition in more precise terms, restricting it to elementary systems. Let then $\mathfrak{a}$ and $\mathfrak{b}$ be such systems.

DEFINITION 1.4.2. *We shall say that the history of* $\mathfrak{a}$ coincides with that of $\mathfrak{b}$ up to a translation in time $t \to t+r$, $\mathfrak{a} =_\tau \mathfrak{b}$, *if and only if the following two conditions are satisfied*:

(i) *the translation* $t \to t+r$ *maps* $T(\mathfrak{a})$ *onto* $T(\mathfrak{b})$, *i.e.* t *is in* $T(\mathfrak{a})$ *if and only if* $t+r$ *is in* $T(\mathfrak{b})$,

(ii) *for each* t *in* $T(\mathfrak{a})$, $\mathfrak{a}(t) = \mathfrak{b}(t+r)$.

In what follows, when histories of two systems coincide up to a translation in time we shall often say that they coincide. By the history of $\mathfrak{a}$ in u, where $u \subseteq T(\mathfrak{a})$, we shall mean the history of the restriction $\mathfrak{a}/u$ of $\mathfrak{a}$ to u. Thus we shall say that the history of $\mathfrak{a}$ in u coincides with that of $\mathfrak{b}$ in v if and only if the histories of $\mathfrak{a}/u$ and $\mathfrak{b}/v$ coincide up to a translation τ, $\mathfrak{a}/u =_\tau \mathfrak{b}/v$. Somewhat incorrectly we shall sometimes write $\mathfrak{a}/u = \mathfrak{b}/v$.

By a *possible history* of $\mathfrak{A}$ we shall understand the history of a variant $\mathfrak{B}$ of that system.

1.5. COMPLEX SYSTEMS

Although the concepts we are going to consider now can be defined in a more general manner, we shall restrict our discussion to elementary systems only. Thus throughout this Section by a system we shall mean an elementary system.

The way in which a system behaves may depend on interaction between the system and its environment. In general then, when trying to describe a system $\mathfrak{a}$, we should not restrict our interest to the variables by means of which the state of the system $\mathfrak{a}$ is characterized but we ought to take into consideration the parameters which give an account of the state of the environment as well.

From the formal point of view instead of studying two systems

$$\mathfrak{a}' = (F'_1, \ldots, F'_k),$$
$$\mathfrak{a}'' = (F''_1, \ldots, F''_l),$$

$\mathfrak{a}'$ being an environment of $\mathfrak{a}''$, it is more practical to combine them together and to form in this way a single system

$$\mathfrak{a} = (F'_1, \ldots, F'_k, F''_1, \ldots, F''_l).$$

In order to indicate how he system $\mathfrak{a}$ has been formed we shall sometimes write:

$$\mathfrak{a} = \langle \mathfrak{a}', \mathfrak{a}'' \rangle.$$

Similar notation will be applied when $\mathfrak{a}$ results from combining more than two systems.

The variables F'_i of $\mathfrak{a}$ are called *input variables* or *inputs of* $\mathfrak{a}$ and the variables F''_i are called *output variables*, or *outputs* of that system, especially if their changes affect environment. Accordingly the state $\mathfrak{a}'(t)$ of $\mathfrak{a}'$ at time t will be called the *input-state* of $\mathfrak{a}$ at time t, and similarly the state $\mathfrak{a}''(t)$ will be referred to as the *output-state* of $\mathfrak{a}$ at time t. We shall call a system whose variables are divided into inputs and outputs an *open system*, or else a *closed* one.

Given two open systems $\mathfrak{a}$ and $\mathfrak{b}$, we shall say that an output F of $\mathfrak{a}$ is *coupled* with an input G of $\mathfrak{b}$ within $u \subseteq T(\mathfrak{a}) \cap T(\mathfrak{b})$ if and only if

(1) $$F(t) = G(t),$$

for all t in u.

We shall say that $\mathfrak{a}$ is *coupled* with $\mathfrak{b}$ if and only if an output of $\mathfrak{a}$ is coupled with an input of $\mathfrak{b}$. Observe that the relation of being coupled is not symmetric, i.e. the fact that $\mathfrak{a}$ is coupled with $\mathfrak{b}$ need not imply that $\mathfrak{b}$ is coupled with $\mathfrak{a}$.

If $\mathfrak{a}$ is coupled with $\mathfrak{b}$ and $\mathfrak{b}$ is coupled with $\mathfrak{a}$, the two systems are said to have a *feedback*.

Now we are in a position to define the notion of a complex system. Let $\mathfrak{R}$ be a class of open systems of the same duration.

DEFINITION 1.5.1.

A. *We shall say that* $\mathfrak{R}$ *is a* complex system *if and only if for any two nonempty subsets* $\mathfrak{R}_1, \mathfrak{R}_2$ *of* $\mathfrak{R}$ *such that* $\mathfrak{R}_1 \cup \mathfrak{R}_2 = \mathfrak{R}$, *there are systems* $\mathfrak{a}$ *in* $\mathfrak{R}_1$ *and* $\mathfrak{b}$ *in* $\mathfrak{R}_2$ *such that* $\mathfrak{a}$ *is coupled with* $\mathfrak{b}$ *or* $\mathfrak{b}$ *is coupled with* $\mathfrak{a}$.

B. *If moreover* $\mathfrak{R}$ *includes a closed chain of systems, i.e., a sequence* $\mathfrak{a}_1, \ldots, \mathfrak{a}_n$ *of systems such that each* $\mathfrak{a}_i$ *is coupled with* $\mathfrak{a}_{i+1}$, $i = 1, \ldots, m-1$, *and* $\mathfrak{a}_m$ *is coupled with* $\mathfrak{a}_1$, *then* $\mathfrak{R}$ *is said* to have a feedback.

It is easily seen that if a complex system $\mathfrak{R}$ is finite, then it can be presented as an elementary system

$$\mathfrak{a}_{\mathfrak{R}} = \langle \mathfrak{a}_1, \mathfrak{a}_2, \ldots, \mathfrak{a}_n \rangle,$$

where $\mathfrak{a}_1, \ldots, \mathfrak{a}_n$ are all elements of $\mathfrak{R}$ and $\mathfrak{a}_{\mathfrak{R}}$ is the system which comprises all the variables of the systems $\mathfrak{a}_i$. The way in which the systems in $\mathfrak{R}$ are coupled is reflected in $\mathfrak{a}_{\mathfrak{R}}$ by the fact that for some pairs of the variables of that system the identities analogous to (1) are satisfied.

It may happen that the way in which $\mathfrak{R}$ is visualized makes it easier or, quite to the contrary, more difficult to grasp regularities which govern its behaviour. The fact then that $\mathfrak{R}$ and $\mathfrak{a}_{\mathfrak{R}}$ are in a sense equivalent need not mean that it is of no practical importance whether $\mathfrak{R}$ is given as a set of elementary systems which are combined into a complex one by a network of couplings or as a single elementary system.

1.6. EMPIRICAL PHENOMENA

Whenever we speak about a phenomenon we shall mean a phenomenon which, at least in general, may occur at different places and at different periods of time. We shall then distinguish between a *phenomenon in the general sense* and its *occurrences* or *instances*, the latter to be referred to as *phenomena in the singular sense*. Phenomena in the general sense will be always treated as sets of their instances. Thus for example, the earthquakes in Lisbon in 1875, in Messina in 1908, in Scopje in 1960, are specific occurrences of the same phenomenon "earthquake". Falling down of an object a_1 in the gravitational field of Earth and falling down of an object a_2 are two specific instances of the phenomenon "falling down in the gravitational field of Earth".

The examples we have produced may be misleading. Note that both the earthquake and the falling down in the gravitational field of Earth can be studied under many different aspects or, to put it more precisely, with respect to many different variables. Although we may assume that we deal with the same phenomenon regardless of whether the earthquake is studied as a geological phenomenon or as a disaster, i.e. a certain social phenomenon, it is more convenient to speak in such a case about two different phenomena though interrelated in a certain important way. We shall take the latter standpoint and consequently assume that two phenomena are identical only if they are studied with respect to exactly the same variables.

The notion of a phenomenon involves then unavoidably certain subjective element: a phenomenon, when it becames an object of our inquiry, is not simply a part of empirical reality but a part of an empirical reality studied with the help of a certain conceptual apparatus. The important thing is to realize that a choice of this apparatus, or as we shall occasionally say a *conceptualization* of a phenomenon, plays a twofold role. On one hand it restricts our area of interest and makes it definite, on the other it suggests certain way of conceiving the phenomenon.

Usually, and in particular throughout this chapter and the next one, we shall assume that empirical phenomena can be viewed as systems, or if they are phenomena in the general sense, as sets of systems of the sort we have defined in Section 1.2. More precisely, we shall define a set of systems $\mathscr{P}$ to be a *phenomenon* or an *abstract phenomenon*, as we shall occasionally say, if and only if all systems in $\mathscr{P}$ are pairwise similar. Perhaps it should also be required that $\mathscr{P}$ is *homogeneous* in the sense that there is a system $\mathfrak{A}_{\mathscr{P}}$ (which need not be an element of $\mathscr{P}$) such that all systems in $\mathscr{P}$ are subsystems of $\mathfrak{A}_{\mathscr{P}}$. This would guarantee that given any two systems:

$$\mathfrak{A} = (A, F_1^A, \ldots, F_n^A),$$
$$\mathfrak{B} = (B, F_1^B, \ldots, F_n^B)$$

in $\mathscr{P}$, such that neither the sets A, B nor durations $T(\mathfrak{A}), T(\mathfrak{B})$ are disjoint, any two of the corresponding variables F_i^A, F_i^B must have the same values when they refer to the same arguments. Thus the corresponding variables can be treated as „the same" but restricted to different sets of objects and different intervals of time. The requirement of homogeneity, although perfectly natural, is dropped out from the definition of a phenomenon in order to make formal considerations run smoother.

Time and again, we shall discuss relations holding between phenomena meant to be a part of physical world, and their abstract, set-theoretic representations. However a brief comment on a certain aspect of the problem seems now to be in order. It is rather obvious that both single theorems and theories as a whole can be relatively easily viewed as theorems or respectively theories about abstract structures defined by means of suitable mathematical conditions. The very nature of the concepts which theories, especially mathematized ones, involve is that abstract objects are natural candidates for being their denotations. At the same time the nature of correspondence between theoretic concepts and real empirical phenomena seems to be enigmatic. The real world does not contain such things as massless springs, 0-dimensional particles or ideal gases. No real number can be claimed to be the exact value of temperature in such and such place, there is nothing like an exact position of a physical object or exact Peter's blood pressure, etc.

Such, both simple and obvious, observations seem to provide a convincing justification for the widely spread belief that subjects of theoretic considerations are models of phenomena not the phenomena themselves. I am afraid that the idea of a phenomenon treated like a ghost which can be contacted only with the help of a medium is misleading. It is perfectly obvious that variables such as "absolutely exact position" or "absolutely exact blood pressure" are not uniquely defined when applied to physical objects, but they can appear in the language in which physical objects are studied. Moreover, which is obvious, one can meaningfully utter assertions such as "the position of the ship is such and such within such and such limit of error" or "Peter's blood pressure measured with such and such accuracy is such and such", referring them directly to empirical objects.

In what follows we shall often pretend that phenomena and their abstract representations are the same things. This way of speaking is not only convenient but at the same time motivated by the role the representation of a phenomenon is expected to play.

Let us conclude this Section with a few rather sketchy remarks on the criteria for deciding what should be treated as an instance of a phenomenon. A suitable example may help in discussing the matter. Assume then that we want to establish correlations between the living conditions and the ratio of the cases of a particular disease, say tuberculosis, in a certain population p. Clearly we may study each case of tuberculosis separately but suppose that we are interested in certain statistic regularities characteristic of p as a whole.

Denote the phenomenon we want to deal with as *TB*. We may figure that its conceptualization will comprise some qualitative variables $C_1, \ldots, C_m$ in terms of which the population p can be divided into different categories such as, say: sex, age group, profession, occupation, etc., certain variables $V_1, \ldots, V_n$ in terms of which living conditions of different subpopulations of the population p can be characterized, and above all the variable T standing for "the ratio of tuberculosis in a subpopulation at a given instant of time". We may still enlarge the list of variables by adding, for instance, one which would serve for distinguishing different forms of tuberculosis, but for the purpose of our example it is irrelevant what variables have finally been selected.

Denote as V the set of parameters $C_1, \ldots, C_m, V_1, \ldots, V_m, T$, and denote as TB_V the phenomenon *TB* studied under the conceptualization V. There are no obstacles to identify TB_V with the generic structure

(1) $\mathfrak{p} = (2^p, C_1, \ldots, C_m, F_1, \ldots, F_n, T)$,

where 2^p is the set of all subsets of the population p, and thus to define TB_V to be a singular phenomenon, as we shall call phenomena which comprise exactly one system. Duration $T(\mathfrak{p})$ of $\mathfrak{p}$ should be defined to be the interval of time during which the population p is examined. The same conceptualization V can be used, however, to study another population q or even the same population p but in a different period of time. The corresponding systems, denote them as $\mathfrak{q}$ and $\mathfrak{p}'$, respectively, may be treated as phenomena different from TB_V and different from each other, but we may as well consider them, and perhaps some other systems of a similar kind, as specific cases of TB_V.

The fact that $\mathfrak{p}$, $\mathfrak{p}'$, $\mathfrak{q}$ possess a common conceptualization need not provide a decisive argument in favour of the latter decision. If there are good reasons to suspect that certain parameters which were not taken into account (and hence were not included into V) affect the regularities observed in different systems in drastically different ways, we may prefer to speak, when referring to those systems, about different phenomena rather than different exemplifications of the same phenomenon. The same living conditions (to be more precise: the same up to certain variables) may have different effect on spread of the disease in tropical and in temperate climate, or say in XVI and in XX century. To the contrary, when we see no substantial differences between surrounding conditions in which different populations are studied, and, what is more, the evidence gathered confirms our conjecture that there are no such differences, we have a good reason for considering all the systems examined as different instances of the same phenomenon.

The whole problem we have discussed arises only when one insists that some sets of similar systems do not deserve to be termed "a phenomenon". It would cause much inconvenience however to accept such a standpoint. By a phenomenon, in the technical sense of the word, we shall understand a set of similar systems, no matter what criteria of combining them together have been adopted.

1.7. PHASE-SPACE OF A PHENOMENON AND SOME RELATED TECHNICAL NOTIONS

Given an abstract phenomenon $\mathscr{P}$ we shall define symbols $St(\mathscr{P})$ and $St^*(\mathscr{P})$ as follows:

(1) $$St(\mathscr{P}) = \bigcup\{St(\mathfrak{A}) : \mathfrak{A} \in \mathscr{P}\},$$

and similarly

(2) $$St^*(\mathscr{P}) = \bigcup\{St^*(\mathfrak{A}) : \mathfrak{A} \in \mathscr{P}\}.$$

The set $St(\mathscr{P})$, then, consists of all states of all systems in P, and the set $St^*(\mathscr{P})$ consists of all possible states of those systems. In what follows we shall sometimes refer to elements of $St(\mathscr{P})$ as *states* of $\mathscr{P}$ and to elements *of* $St^*(\mathscr{P})$ as *possible states* of $\mathscr{P}$ thus abbreviating the locutions "a state of an instance of $\mathscr{P}$" and "a possible state of an instance of $\mathscr{P}$", respectively.

Observe that if $\mathscr{P}$ is composed of elementary systems, each state $\mathbf{x} = (x_1, \ldots, x_n)$ in $St^*(\mathscr{P})$ determines a point in an n-dimensional Euclidean real space $\mathscr{E}^n$. Given then an abstract phenomenon $\mathscr{P}$ which consists of elementary systems involving n variables, we can represent both $St^*(\mathscr{P})$ and $St(\mathscr{P})$ as certain regions in the space $\mathscr{E}^n$ usually referred to as the *state-space* of the phenomenon.

In turn, if for some $\mathfrak{a}$ in $\mathscr{P}$ $\mathfrak{a}(t) = \mathbf{x}$, the state of the system at time t can be represented by a point (or equivalently: vector) in an $n+1$-dimensional space $\mathscr{E}^{n+1}$, $t, x_1, \ldots, x_n$ being coordinates of that point (vector). Consequently the history of $\mathfrak{a}$ is represented by a curve which is sometimes (cf. e.g. W. R. Ashby, 1960) called the *line of behaviour* of $\mathfrak{a}$. Finally the phenomenon $\mathscr{P}$, as a whole, is represented by the collection of lines of behaviour of the systems it includes. This simple and quite obvious geometrical interpretation is helpful when transformations of states of systems in $\mathscr{P}$ are discussed.

In what follows the space $\mathscr{E}^{n+1}$ will be referred to as the *phase-space* of $\mathscr{P}$, provided that $\mathscr{P}$ is as described above, i.e. it consists of elementary systems involving n-variables. Points of a phase-space will be referred to as *events*. In this way we "decompose" histories (or lines of behaviour) of elementary systems on to events, which may happen to differ only as to "the time of their occurrence", i.e. the value of their time coordinates.

1.8. SEMI-INTERPRETED LANGUAGES

A very general character of our considerations makes it impossible to define the notion of a language in the way which is rigorous enough and at the same time is not too restrictive for our purposes. On the other hand, the notion of a language as will be applied in this book will not involve any peculiarities which might deserve our special attention. We shall confine ourselves to a short and informal survey of: (1) the most important properties of the languages considered in the subsequent parts of the book and (2) the terminology we shall employ speaking about languages and their properties. The preliminary remarks of this section will be supplemented by the discussion to be found in Chapter III.

As it is customary, we shall assume that in order to define a language L it is necessary to define:

(i) the *vocabulary* V of the language $\mathcal{L}$, i.e. the set of objects called *terms* or *elementary expressions* (*symbols*) of $\mathcal{L}$.

(ii) the *rules of formation* of $\mathcal{L}$, i.e. the rules which define the way in which elementary symbols may be concatenated in order to form *compound expressions*, in particular *formulas* and *sentences*.

(iii) the *rules of interpretation* of $\mathcal{L}$,

Moreover the characteristic of the language $\mathcal{L}$ may also involve:

(iv) the *rules of inference* (more generally the rules of constructing proofs).

The set of sentences of $\mathcal{L}$ will be again denoted as $\mathcal{L}$. The ambiguity to which this convention gives rise should not lead to confusions; the context should always make it clear in which sense the symbol $\mathcal{L}$ is applied.

A definition of a vocabulary of $\mathcal{L}$ cannot consist simply in listing the elementary expresions of the language since it is expected to provide also a division of those expressions into *syntactical categories.* Two expressions are said to belong to the same *syntactical category* if they play the same role with respect to formation of compound expressions. More precisely, two expressions e_1, e_2 are of the same *syntactical category* provided that the set of all formulas is closed under replacing one of these expressions by another in all formulas in which they occur.

Apart from being divided into syntactical categories the elementary expressions must be also divided into *constant* and *variable* symbols. Let us call the reader's attention once again to the fact that the linguistic concept of a variable does not coincide with the concept of a system variable. In fact symbols for system variables, say: blood pressure, the amount of goods of specific kind, distance, etc., are constant symbols from the linguistic point of view.

The main task of the rules of formation is to describe the way in which formulas, in particular sentences, can be constructed. A *sentence* is meant to be a formula which does not involve free variables. This definition is based on the fact that any occurrence of a language variable in an expression is either *free* or *bound.* An expression which involves no free variable is called *closed*, or else *open.* All these concepts are assumed to be known; in fact we make such an assumption about almost all concepts discussed in this section. Clearly, for each specific language $\mathscr{L}$ the notions of a free and bound variable as well as those of a closed and open expression (in particular a formula) should be defined by the rules of formation for that language.

The role of the rules of interpretation for the language $\mathscr{L}$ is to define a class of functions referred to as *possible interpretations* for $\mathscr{L}$ (*interpretations* for $\mathscr{L}$, for short). Each interpretation I will be assumed to assign to all terms of $\mathscr{L}$, to all sentences of $\mathscr{L}$, and perhaps also to some of the remaining expressions of $\mathscr{L}$ objects called *values* of those expressions. We shall assume that any two interpretations I, I' for L which coincide on V (i.e. $I(e) = I'(e)$ for all e in V) are identical, i.e. they assign the same value to all expressions on which they are defined.

If e is a constant symbol, the value $I(e)$ of e will be called the *denotation* of e under the interpretation I. The value $I(e)$

of a variable e will be meant to be its *range*, i.e. the set of objects which e *represents* or, in other words, *runs over*. Finally, we assume that each sentence α may take only two values: *truth* or *falsity*. denoted as 1 and 0, respectively. 1 and 0 will usually be referred to as the *truth-values*.

When the interpretation I is fixed we shall often refer to $I(e)$ as the *interpretation* of e or, informally, as the *meaning* of e.

Under certain assumptions each interpretation I for $\mathscr{L}$ may happen to correspond uniquely to a structure of the form

(1) $$(I(e_1), \ldots, I(e_k)),$$

where $e_1, e_2, \ldots, e_k$ are terms of $\mathscr{L}$. If that situation takes place, the structure (1) is often referred to as a *model structure* of I. Usually we shall not discern between an interpretation and its model structure.

The languages with which we are going to deal will be languages of some theories, either mathematical or empirical. The terms such languages involve may be divided into two classes: *auxiliary* and *specific* for a given theory. Loosely speaking the specific terms of a theory Θ are those which refer directly to the objects examined in a given theory. Thus if Θ is a theory of rational numbers, then symbols of rational numbers and symbols of operations on rational numbers are specific for Θ. If Θ is an empirical theory which refers to systems of the form $(A, F_1, \ldots, F_n)$, then, clearly, symbols of the variables F_i are specific terms of Θ. Moreover Θ may involve as specific terms names for some elements of A, variables running over A, time variables and perhaps some other symbols.

The borderline between the specific and the auxiliary terms must be, at least to some extent, arbitrary. The theorem which states that $(Ra, +)$ is a group (Ra being the set of rational numbers, and $+$ the familiar operation of adding numbers) involves the algebraic concept of a group which should be treated as auxiliary rather than specific for the theory of rational numbers. The terms defined by means of specific symbols of the theory may be treated depending on what is more convenient either as specific again or as auxiliary.

In what follows we shall assume that in the case of empirical theories the partition of elementary terms into auxiliary and specific coincides in principle with that into logico-mathematical, and the remaining ones, though it may happen that some mathematical terms (e.g. probability measure) acquire an empirical interpretation on the ground of the theory and thus fall under the category of specific terms. As far as logical terms are concerned we shall assume that the logical apparatus for all the languages is that of the first order predicate calculus with identity.

Whenever a language $\mathscr{L}$ is dealt with we shall assume that among all of its possible interpretations some are distinguished as *standard*. Loosely speaking these are interpretations under which logical and mathematical terms acquire their "standard meaning". Thus for instance, under the standard interpretation, the symbol of disjunction $\vee$ dinotes the operation $\vee$ defined in the known manner $1 \vee 1 = 1 \vee 0 = 0 \vee 1 = 1$, $0 \vee 0 = 0$. The arithmetical relation $\leqslant$, when interpreted in the standard way, holds between 5 and 7 and does not hold between 7 and 5, clearly with the proviso that the interpretation for 5 and 7 is standard also. Those casual remarks give only a rudimentary idea of what is to be meant by a standard interpretation and therefore we shall supplement them in due course by some additional comments and examples (cf. Chapter III, especially Section 3.2).

In what follows the class of all standard interpretations for $\mathscr{L}$ will be denoted as $Int(\mathscr{L})$. The languages for which standard interpretations are defined are sometimes referred to as *semi-interpreted* (cf. e.g. B. van Fraassen, 1970); if this terminology is adopted, a language is meant to be *interpreted* only if from all standard interpretations a certain specific one is singled out and declared to be intended. All the languages which we are going to consider will be semi-interpreted in the sense defined above.

In order to define a standard interpretation for a semi-interpreted language $\mathscr{L}$ it is often enough to define the interpretation for the specific terms of $\mathscr{L}$ (of the theory stated in $\mathscr{L}$). Given an interpretation I for $\mathscr{L}$ call the restriction I/V_0 of I to the specific terms of $\mathscr{L}$ a *partial interpretation* for $\mathscr{L}$.

Let us list some assumptions concerning the languages which will be considered. Some of the assumptions have been already stated. Given any language $\mathscr{L}$ we shall postulate that:

(L1) For any two interpretations I, I' for L, if $I/V = I'/V$ then $I = I'$.

(L2) Among all interpretations for $\mathscr{L}$ some are distinguished as standard.

(L3) Among all elementary expressions of $\mathscr{L}$ some are distinguished as specific for $\mathscr{L}$.

(L4) If, I, I' are standard interpretations for $\mathscr{L}$ and their restrictions to the terms specific for $\mathscr{L}$ coincide, then $I = I'$.

(L5) If I, I' are two isomorphic interpretations for $\mathscr{L}$ (i.e. models corresponding to I and I' are isomorphic), then for each $\alpha \in \mathscr{L}$, $I(\alpha) = I'(\alpha)$.

If furthermore $\mathscr{L}$ satisfies the following two conditions:

(L6) The vocabulary V of $\mathscr{L}$ is effectively definable.

(L7) The set of all sentences of $\mathscr{L}$ is effectively definable. Then we shall call $\mathscr{L}$ *standard.*

We have defined the notion of a standard language in a loose way. Still all conditions the definition involves, but perhaps (L5), might be easily spelled out in a rigorous manner For that purpose it is enough to define a standard language to be a structure of the form

$$\mathscr{L} = (V, V_0, \mathscr{L}, Int^*(\mathscr{L}), Int(\mathscr{L}))$$

consisting of certain abstract objects satisfying certain conditions. The set V would be interpreted as the vocabulary of $\mathscr{L}$, V_0 as the sets of specific terms of $\mathscr{L}$, $\mathscr{L}$ as the set of all sentences of $\mathscr{L}$ (in accordance with the convention adopted earlier we apply the ambiguous notation), $Int^*(\mathscr{L})$ as the class of all possible interpretations for $\mathscr{L}$, finally $Int(\mathscr{L})$ as the class of all standard interpretations for $\mathscr{L}$. Elements of V, V_0, L, $Int^*(\mathscr{L})$, $Int(\mathscr{L})$ should be assumed to satisfy the conditions we have imposed on the language constituents they represent and in particular conditions (L1)-(L7). Thus, for instance we would require $Int^*(\mathscr{L})$ to be the class of functions of a suitable sort and we shall demand that $Int(\mathscr{L}) \subseteq Int^*(\mathscr{L})$. The latter requirement is a formal counterpart of (L2).

There are two concepts involved in the definition of a standard language which deserve special attention. These are effective definability and isomorphism of structures.

Whenever we say that a set is effectively definable, we shall mean that there exists a procedure which is at least recursive (in all actual cases it will be primitive recursive) by means of which elements of that set can be constructed. The recursiveness and also primitive recursiveness are technical concepts whose definitions can be found in text-books on mathematical logic (more specifically on recursion theory) cf. for instance A. Grzegorczyk (1974), or J. R. Schoenfield (1967). In less rigorous parts of the discussion we shall prefer to use the word "effective" rather than recursive.

The concept of isomorphism is fundamental for any mathematical considerations in which structures of any sort whatsoever are involved. Various specifications of that concept conforming to structures of a certain special kind are no doubt very well known to the reader. Incidentally in Section 1.2 we have defined the isomorphism of two numerical systems. Still an attempt to define the notion of isomorphism in a fully general manner (so that the concept would be applicable to any couple of structures whatsoever) would most likely result in a fairly complicated and difficult to grasp definition. More practical policy then is to define the concept whenever there is a need for it for any considered type of structures separately.

Let us conclude our remarks concerning the notion of a language with a few comments on the notion of an intended interpretation. It is usually taken for granted that all actually applied (i.e. not artificial) languages possess a certain distinguished and fixed interpretation called *intended*. The meaning which the terms of the language acquire under this interpretation is simply the very meaning in which those terms are applied.

A closer examination of empirical theories reveals that although possible, it is very inconvenient to view the languages of those theories as possessing a fixed intended interpretation. The reason for that is quite simple. Each empirical theory provides an account of regularities characteristic of a class of singular phenomena thus referring simultaneously to all of them. Each singular phenomenon considered under the conceptualization offered by the language of the theory can be viewed as a model for at least some of the concepts of the theory. Different instances of the phenomenon examined (different singular phenomena) may correspond then to different interpretations of the language, all of which deserve to be called intended! The examples we shall discuss later (starting already with some parts of Chapter II) should throw more light on the idea of multiple referential languages to which the comments presented above amount.

NOTE. Although many researchers in the field of methodology and philosophy of science take it for granted that empirical theories should be treated as related to a class of empirical systems rather than to a single system, this approach is sometimes questioned. The main argument raised against the idea of multiple reference can be summarized in a few words: each empirical theory is related to a certain "part" of the empirical world and this part is the unique and proper interpretation for the theory.

The idea of a single interpretation which represents empirical world (or a part of it) is in an explicit form involved in Tarski's considerations concerning the notion of truth, (cf. A. Tarski, 1933; 1944). It has been pursued further by many authors in a formal or semi-formal way, (cf. M. Kokoszyńska, 1936, R. Carnap, 1942; 1966, H. Mehlberg, 1958, R. Suszko, 1968, M. Przełęcki, 1969).

The multi-referential approach has been suggested by E. Beth, 1949; 1960. It underlies the account of empirical theories adopted by R. Montague (1962), B. van Fraassen (1970); (1972), J. Sneed (1971), F. Suppe (1971), W. Stegmüller (1973). It is also intrinsically incorporated in P. Suppes's method of defining empirical theories as set theoretic predicates, cf. e.g. the definition given in his *Introduction to Logic* on p. 294 which starts: "A system $\mathfrak{p} = P, T, s, m, f, s\rangle$ is a system of particle mechanics if and only if the following seven axioms are satisfied ...".

1.9. FUNDAMENTAL SEMANTIC CONCEPTS

The notion of a standard interpretation for $\mathscr{L}$, although defined in a very general way, suffices to set up the fundamentals of logical semantics. The notions we are going to define belong to the standard equipment of any semantic analysis carried out with logical tools, and thus they should be familiar to most of the readers of that book. Still it would be inappropriate to omit the definitions, if not for any other reason, then because of some notational and terminological details which should be brought out.

If α is a sentence of $\mathscr{L}$ and for some interpretation I, $I(\alpha) = 1$, we shall say that α is *valid under the interpretation I*, or equivalently is *valid in the model* structure I (model for $\mathscr{L}$ determined by I). The set of all sentences of $\mathscr{L}$ valid under I will be denoted as

$$Val(I).$$

Instead of $Val(I)$ we shall occasionally write also Val_I.

An interpretation I will be said to be a *realization* for $X \subseteq \mathscr{L}$ or equivalently a *model* for X if and only if $X \subseteq Val(I)$. The class of all standard realizations for X, i.e. realizations being standard interpretations for $\mathscr{L}$ will be denoted as

$$Mod(X).$$

If $Mod(\alpha) = Int(\mathscr{L})$, i.e. each standard interpretation for $\mathscr{L}$ is a realization for the sentence α (the unit set $\{\alpha\}$), α will be said to be a *mathematical identity*. The adjective "mathematical" is to call attention to the fact that as a rule the validity of a sentence being a mathematical identity under all standard interpretations is due to the meaning of mathematical and not only logical terms the sentence involves. Mathematical identities, are for instance the sentences: "For all φ, $\sin^2\varphi + \cos^2\varphi = 1$", $5 \leqslant 7$, $dx^2/dx = 2x$, whose validity cannot be established by means of purely logical considerations.

Clearly, some of mathematical identities are logical *tautologies*, e.g. "For all x either $x \leqslant 2$ or not $x \leqslant 2$", i.e. sentences valid (in all standard interpretations) in virtue of the meaning of logical terms only.

If $Mod(X) \subseteq Mod(\alpha)$, the sentence α will be said to be *entailed by X*, in symbols

$$X \models \alpha.$$

If $X \models \alpha$ does not hold, we shall occasionally write

$$X \not\models \alpha.$$

Instead of $\emptyset \models \alpha$ we shall usually write

$$\models \alpha.$$

The reader can easily verify that $\models \alpha$ if and only if α is a mathematical identity.

Given a set of sentences X we shall say that it is *semantically consistent* if and only if X has a realization, or else we shall say that X is *semantically inconsistent*. We clearly have the following

ASSERTION 1.9.1. *X is semantically inconsistent if and only if for each α, $X \models \alpha$.*

Two sets sentences X and Y will be said to be *semantically equivalent* if and only if $Mod(X) = Mod(Y)$. Note that

ASSERTION 1.9.2. *X and Y are semantically equivalent if and only if for each α,*

$$X \models \alpha \text{ if and only if } Y \models \alpha.$$

In particular any two inconsistent sets and any two sets of tautologies are semantically equivalent.

Write $\alpha \approx \beta$ whenever α and β are semantically equivalent. It is immediately seen that $\alpha \approx \beta$ if and only if $\models \alpha \Leftrightarrow \beta$, where $\Leftrightarrow$ is the equivalence connective.

If for any α in X, $X - \{\alpha\} \not\models \alpha$, sentences in the set X will be said to be *semantically independent*.

Let me end this short survey of semantic notions with a few remarks on the links between the concept of *validity* and that of *truth*. These two concepts will be applied in the following way. Whenever α is valid under I we may as well say that α is *true* under I. But we shall prefer to limit applications of the term "true" to those interpretations only which are intended. Thus whenever α is said to be true under I it will be understood that I is an intended interpretation for the language considered and α is valid under I. A similar convention will be adopted for the term "false". α will be said to be *false* under I whenever

I is an intended interpretation for the language to which α belongs and α is not valid in I.

Finally we should perhaps comment briefly on the assumption we have made that the languages we are going to consider can be viewed as such that for each α and for each model I, α is either valid in I or not. The dychotomy presupposed by this assumption rules out the possibility to consider as well-formed formulas expressions like $k \sin^{-1} \varphi = l \cos^{-1} \varphi$, where either $\sin \varphi$ or $\cos \varphi$ equals 0, or like $\frac{d}{dx} f(x) = g(x)$, where f is not a continuous function. This conforms to common practice, but at the same time gives rise to certain nontrivial difficulty. It may happen to be necessary to define a sentence (more generally a well-formed formula) relative to a particular interpretation for the symbols involved in the formula considered. Thus e.g. $\frac{d}{dx} f(x) = g(x)$ may happen to be a well-formed formula under some interpretations and a meaningless expression under others. Consequently an effective definition of the set of sentences may become unavailable.

There are several ways in which the mentioned difficulty can be solved. Although it certainly deserves some attention we shall not discuss the matter since it concerns certain technicalities of minor importance for our further considerations.

1.10. FORMALIZED LANGUAGES AND DEDUCTIVE CONCEPT OF A THEORY

In order to establish that X entails α, a proof of α from X must be given. If the notion of proof has been defined, the language $\mathscr{L}$ will be said to be *formalized*.

In what follows by a *logical proof* from X we shall always mean a finite sequence of sentences

(1) $\alpha_1, \alpha_2, \ldots, \alpha_k$

such that each α_i in the sequence is either an element of X

or results from some of the sentences preceding it in the sequence by a logical rule of inference.

If a proof from X of the form (1) is given, the elements of $X \cap \{\alpha_1, \ldots, \alpha_k\}$ are said to be the *premises* of the proof and α_k the *conclusion*.

The logical notion of a proof is however somewhat too restrictive for our purposes. Given a language $\mathscr{L}$, we shall assume that some of the mathematical identities can be applied in the same way as the premises, i.e. we define α to be *provable* from X if there exists a logical proof of α from X and the mathematical identities we have distinguished.

The reason why all mathematical identities should not be admitted in proof is obvious. The set of all proofs should be *effectively definable*, i.e. there should exist an effective procedure which allows us to decide whether a given sequence is a proof or not. The set of all mathematical identities is not effectively definable (already the set of logical tautologies is not effectively definable) and thus were all mathematical identities admitted in proof, the set of all proofs would not be effectively definable either.

The reason why we must insist on decidability of the notion of a proof is clear. If there were no effective procedures by means of which one could verify whether an alleged proof is a proof indeed, disagreement among specialists concerning correctness of such a proof would not be decidable in a commonly accepted, intersubjective manner.
cepted, intersubjective manner.

Applying the well known notation we shall write

$$X \vdash \alpha$$

whenever there is a proof of α from X (α is *derivable* from X) or else

$$X \nvdash \alpha.$$

The notation $\varnothing \vdash \alpha$ will be abbreviated to $\vdash \alpha$. A sentence α such that $\vdash \alpha$ will be referred to as *mathematically provable*.

The way in which $\vdash$ has been defined guarantees that the derivability relation is *sound*, i.e. the following holds true:

ASSERTION 1.10.1. *For all X and α (of a fixed language $\mathscr{L}$) $X \models \alpha$ whenever $X \vdash \alpha$.*

One may easily see that in general the converse is not true. Consider for instance the set of all intervals of the form $\left(0, \frac{1}{k}\right]$, where $k = 1, 2, \ldots$, and assuming that A denotes a set, form an infinite set of sentences X consisting of all sentences of the form

$$A \subseteq \left(0, \frac{1}{k}\right].$$

Since $\bigcap_k \left(0, \frac{1}{k}\right] = \varnothing$, we immediately conclude that $A = \varnothing$. We have then

$$X \models A = \varnothing$$

but at the same time

$$X \nvdash A = \varnothing.$$

Indeed, were A derivable from X, it would be derivable from a finite subset of X, because all proofs are assumed to be finite. But one may easily verify that for all finite subsets $X_f \subseteq X$

$$X_f \not\models A = \varnothing.$$

Thus also, by the soundness of $\vdash$ (cf. Assertion 1.10.1)

$$X_f \nvdash A = \varnothing.$$

The following is easily seen to be valid.

ASSERTION 1.10.2. *For all $X, Y \subseteq \mathscr{L}$ and for all $\alpha \in \mathscr{L}$:*

(i) *If $\alpha \in X$, then $X \vdash \alpha$.*
(ii) *If $\{\alpha : X \vdash \alpha\} \vdash \beta$, then $X \vdash \beta$.*
(iii) *If $X \subseteq Y$ and $X \vdash \alpha$, then $Y \vdash \alpha$.*
(iv) *If $X \vdash \alpha$, then for some finite $X_f \subseteq X$, $X_f \vdash \alpha$.*

Incidentally, we have already made use of (iv) in the argument which shows that $\models$ and $\vdash$ do not coincide. We shall refer to the property of $\vdash$ asserted by (iv) by saying that $\vdash$ is *finitary*.

The next assertion is more sophisticated, still we shall not present the proof of it.

ASSERTION 1.10.3. *For all* $X \subseteq \mathscr{L}, \alpha_1, \ldots, \alpha_k, \beta \in \mathscr{L}$, *the following conditions hold true*:

(v) $X \cup \{\alpha_1, \ldots, \alpha_k\} \vdash \beta$ *if and only if* $X \vdash (\alpha_1 \wedge \alpha_2 \wedge \ldots \ldots \wedge \alpha_k) \rightarrow \beta$.

(vi) $\{\alpha, \neg \alpha\} \vdash \beta$.

(vii) $X \cup \{\neg \alpha\} \vdash \alpha$ *if and only if* $X \vdash \alpha$.

(viii) $X \cup \{\alpha \vee \beta\} \vdash \gamma$ *if and only if* $X \cup \{\alpha\} \vdash \gamma$ and $X \cup \cup \{\beta\} \vdash \gamma$.

The "if" part of Condition (v) (the implication from left to right) is known as *Deduction Theorem*. Conditions (vi), (vii) provide a characteristic of properties of the negation sign and disjunction sign in terms of derivability relation.

The counterparts of semantic concepts defined in terms of derivability relation instead of entailment will be often referred to as *deductive*. Thus if for each β, $X \vdash \beta$, X will be said to be *deductively inconsistent* or else *deductively consistent*. Note that by Condition (vi) of Assertion 1.10.3. X is deductively inconsistent if and only if for some α, both $X \vdash \alpha$ and $X \vdash \neg \alpha$.

Two sets X, Y will be said to be *deductively equivalent* if and only if for each α:

$$X \vdash \alpha \text{ if and only if } Y \vdash \alpha.$$

We shall indicate the deductive equivalence of two sentences α, β by writing

$$\alpha \sim \beta$$

One may easily see that $\alpha \sim \beta$ whenever $\vdash \alpha \leftrightarrow \beta$.

A set X will be said to be *deductively independent* whenever for any α in X,

$$X - \{\alpha\} \nvdash \alpha$$

Any finite set X_f deductively equivalent to X will be called an *axiomatics* (a set of *axioms*) for X. It is convenient to extend the notion of an axiomatics so that it covers also infinite equivalents to X provided that they are recursively definable. Note that we do not demand an axiomatics to be a deductively independent set.

A set X will be said to be *deductively closed* whenever $X \vdash \alpha$ implies that $\alpha \in X$.

The last two concepts which we have introduced are not counterparts of any semantic concepts defined earlier, though obviously such counterparts can be easily defined.

Let us examine closer how the concepts of a semantically consistent and a deductively consistent sets of sentences are related to each other.

Obviously, whenever X is semantically consistent it is deductively consistent. In fact the construction of a model in which X is valid is one of the main methods of proving consistency of X. The converse in general need not be valid. Consider the same set of sentences X we constructed in order to prove that $\models$ and $\vdash$ do not coincide, i.e. the set of all sentences of the form

$$A \subseteq \left(0, \frac{1}{k}\right],$$

and enlarge X by adding to it the sentence $A \neq \varnothing$. The resulting set X' is semantically inconsistent since $X \models A = \varnothing$, but it is deductively consistent since each finite subset of X' is (as easily seen) consistent. We obviously have

Assertion 1.10.4. *X is deductively inconsistent if and only if for some finite $X_f \subseteq X$, X_f is inconsistent.*

Proof. Clearly if X_f is inconsistent, then by (iii) in Assertion 1.10.2 X must be inconsistent.

In turn suppose that X is inconsistent. Then for some α, $X \vdash \alpha$ and $X \vdash \neg \alpha$ or equivalently $X \vdash \alpha \wedge \neg \alpha$. But $\vdash$ is finitary and hence for some finite X_f, $X_f \vdash \alpha \wedge \neg \alpha$, which concludes the proof.

1.11. A CRITERION OF CONSISTENCY

We shall state quite a sophisticated criterion of consistency of a set of sentences which will be of some significance for further considerations. Before we do that we must define a few algebraic notions which all by themselves are of considerable importance.

DEFINITION 1.11.1. *Given a set A and a binary relation $\leqslant$ defined on it we shall say that $\leqslant$ is an* ordering *on A, and $(A, \leqslant)$ is an* ordered set *if and only if $\leqslant$ is reflexive, transitive, and anti-symmetric, i.e. for all x, y, z in A the following conditions hold true*:

(i) $x \leqslant x$.

(ii) *If* $x \leqslant y$, $y \leqslant z$, *then* $x \leqslant z$.

(iii) *If* $x \leqslant y$ *and* $y \leqslant x$, *then* $x = y$.

An element a of an ordered set $(A, \leqslant)$ is said to be *maximal* (*minimal*) if there is no element b in A such that $a \leqslant b$ ($b \leqslant a$) and $a \neq b$. An ordered set can have several maximal or minimal elements.

An element a of an ordered set $(A, \leqslant)$ is said to be the *greatest* (*least*) one if for every $x \in A$, $x \leqslant a$ ($a \leqslant x$). It follows from this definition and (iii) that an ordered set can have at most one greatest element and at most one least element. The greatest (least) element of an ordered set, if it exists, will be denoted by 1_A (by 0_A).

An element a of an ordered set $(A, \leqslant)$ is said to be an *upper* (*lower*) *bound* of a non-empty subset A_0 of A if $b \leqslant a$ ($a \leqslant b$) for every $b \in A_0$. If the set of all upper (lower) bounds of A_0 contains a least (greatest) element, then this element is called *supremum of* A_0, *sup* A_0 (*infimum of* A_0, *inf* A_0) and is denoted by $\bigcup A_0$ ($\bigcap A_0$). It follows from this definition that $\bigcup A_0 = a$ ($\bigcap A_0 = a$) if and only if the following conditions are satisfied:

(1) $\qquad b \leqslant a$ ($a \leqslant b$) for every $b \in A_0$,

(2) if $c \in A$ and $b \leqslant c$ ($c \leqslant b$) for every $b \in A_0$, then $a \leqslant c$ ($c \leqslant a$).

An ordering relation $\leqslant$ on a set A is called a *linear ordering* if the following condition is satisfied:

(iv) $x \leqslant y$ or $y \leqslant x$, for every $x, y \in A$.

A *chain* is a pair $(A, \leqslant)$, where A is a non-empty set and $\leqslant$ is a linear ordering on A. In every chain the notions of a greatest (least) element and a maximal (minimal) element coincide. Observe that if $\leqslant$ is a linear ordering on A and $A_0 \subseteq A$, then the relation $\leqslant$ restricted to A_0 is a linear ordering on A_0.

We quote without proof the following well known theorem.

KURATOWSKI-ZORN LEMMA. *If $(A, \leqslant)$ is an ordered set and every chain $(B, \leqslant|_B)$, where $B \subseteq A$, has an upper bound in A, then for each $a_0 \in A$ there exists a maximal element a in A such that $a_0 \leqslant a$.*

DEFINITION 1.11.2.

A. *A structure $(A, \cup, \cap)$ with two binary operations on a non-empty set A will be said to be* a lattice *if and only if*:

(i) *The binary relation $\leqslant$ defined on A by the condition $a \leqslant b$ if and only if $a \cup b = b$ is an ordering on A.*

(ii) *For any a, b in A, $a \cup b$ is $\sup \{a, b\}$ and $a \cap b$ is $\inf \{a, b\}$ with respect to $\leqslant$.*

B. *If moreover the following two conditions are satisfied*:

(iii) $a \cap (b \cup c) = (a \cap b) \cup (a \cap c)$,

(iv) $a \cup (b \cap c) = (a \cup b) \cap (a \cup c)$,

the lattice will be said to be distributive.

We are now in a position to define the main algebraic concept we need.

DEFINITION 1.11.3.

A. *A structure $\mathscr{A} = (A, \cup, \cap, 1, 0)$ will be said to be a* Boolean algebra *if and only if the reduct $(A, \cup, \cap)$ of it is a distributive lattice with 1 and 0 being its greatest and least elements, respectively.*

B. *If moreover for each infinite sequence $a_1, a_2, \ldots$ of elements of A there exist both supremum and infimum of the set of elements of that* *sequence (we denote them as* $\bigcup a_n$ *and* $\bigcap a_n$ *respectively), then* $\mathscr{A}$ *will be said to be a* Boolean σ-algebra.

The elements 1 and 0 of $\mathscr{A}$ (which will be occasionally denoted also as $1_{\mathscr{A}}$ and $0_{\mathscr{A}}$) are called the unit and zero elements of $\mathscr{A}$. One may easily see that if $\mathscr{A}$ contains more than one element, then $1 \neq 0$. If $1 = 0$, the algebra $\mathscr{A}$ is called *trivial*.

As known, if $\mathscr{A}$ is a Boolean algebra, then for each element a of $\mathscr{A}$ there exists exactly one element denoted as $-a$ such that

$$a \cup -a = 1$$

and

$$a \cap -a = 0.$$

The element $-a$ is called the *complement* of *a*, and $-$ is a unary operation on A.

DEFINITION 1.11.4.

A. *Given two Boolean algebras* $\mathscr{A}$, $\mathscr{B}$, *A and B being the sets of their elements respectively, and given a mapping* $h: A \to B$, *we shall say that h is a* Boolean homomorphism *from* $\mathscr{A}$ *into* $\mathscr{B}$ *if and only if for all* a, b in A:

(H1) $h(a \cup b) = ha \cup hb$.
(H2) $h(a \cap b) = ha \cap hb$.
(H3) $h1_{\mathscr{A}} = 1_{\mathscr{B}}$.
(H4) $h0_{\mathscr{A}} = 0_{\mathscr{B}}$.

B. If *moreover for each infinite sequence* $a_1, a_2, \ldots$ *of elements of* $\mathscr{A}$,

(σH1) *whenever* $\bigcup a_n$ *exists, then* $\bigcup ha_n$ *exists and* $h \bigcup a_n = \bigcup ha_n$, *and similarly*

(σH2) *whenever* $\bigcap a_n$ *exists, then* $\bigcap ha_n$ *exists, and* $h \bigcap a_n = \bigcap ha_n$, *then h is said to be a* Boolean σ-homomorphism.

It is a matter of an easy proof to show that if h is a Boolean homomorphism from $\mathscr{A}$ to $\mathscr{B}$, then for each a in A

(H5) $h(-a) = -ha$.

One may easily verify that the relation $\sim$ of deductive equivalence of sentences is reflexive, transitive, and symmetric, i.e. for all α, β, γ:

(i) $\alpha \sim \alpha$.
(ii) If $\alpha \sim \beta$ and $\beta \sim \gamma$, then $\alpha \sim \gamma$.
(iii) If $\alpha \sim \beta$, then $\beta \sim \alpha$.

Let us state without proof the following known theorem (cf. e.g. Rasiowa, Sikorski, 1963).

THEOREM 1.11.5. *Let* $\mathscr{L}$ *be a language. Define the operations* $\cup, \cap, -$ *on the quotient set* $\mathscr{L}/\sim$ *(i.e. the set of all equivalence classes of the form* $|\alpha| = \{\beta : \alpha \sim \beta\}$*) as follows:*

(i) $|\alpha| \cup |\beta| = |\alpha \vee \beta|$,
(ii) $|\alpha| \cap |\beta| = |\alpha \wedge \beta|$,
(iii) $-|\alpha| = |\neg \alpha|$,

and put

(iv) $1 = |\alpha \vee \neg \alpha|$,
(v) $0 = |\alpha \wedge \neg \alpha|$.

Then the algebra $\mathbf{L} = (\mathscr{L}/\sim, \cup, \cap, -, 1, 0)$ *is a Boolean algebra.*

The reader may easily verify that all constituents of **L** are well defined.

The algebra **L** is known as a *Lindenbaum-Tarski algebra*. The notion of a Lindenbaum-Tarski algebra is one of the basic tools in metamathematical investigations. In particular the following theorem, which is just the criterion to be stated, is valid. We shall quote it without proof.

Denote as **2** the two-element Boolean algebra whose only elements are 1 and 0.

THEOREM 1.11.6. *A set of sentences* $X \subseteq \mathscr{L}$ *is deductively consistent if and only if there exists a Boolean homomorphism* $h: \mathscr{L}/\sim \to \mathbf{2}$*, such that for each* α *in* X, $h\alpha = 1$.

Apart from the notion of a Boolean algebra we shall also need the notion of an algebra of sets. The two notions are related to each other in a well-known manner, namely each Boolean algebra is isomorphic to an algebra of sets (*Stone's Theorem*), on the other hand each algebra of sets is a Boolean algebra.

DEFINITION 1.11.7.

A. $\mathscr{F}$ *is an* algebra of sets on X *if and only if* $\mathscr{F}$ *is a non-empty family of subsets of* X *and for every* A *and* B *in* $\mathscr{F}$:

(i) $A \in F$;

(ii) $A \cup B \in \mathscr{F}$.

B. *Moreover, if* $\mathscr{F}$ *is closed under countable unions, that is, if for* $A_1, A_2, \ldots, A_n \ldots \in F$, $\bigcup_{i=1}^{\infty} A_i \in \mathscr{F}$, *then* $\mathscr{F}$ *is a* σ-algebra on X.

One may easily prove that if $\mathscr{F}$ is an algebra of sets on X then,

(i) $X \in \mathscr{F}$,

(ii) $\varnothing \in \mathscr{F}$,

(iii) If $A \in \mathscr{F}$ and $B \in \mathscr{F}$, then $A \cap B \in \mathscr{F}$,

(iv) If $A \in \mathscr{F}$ and $B \in \mathscr{F}$, then $A - B \in \mathscr{F}$,

(v) If moreover $\mathscr{F}$ is a σ-algebra as well, and $A_1, A_2, \ldots, A_n \ldots \ldots \in \mathscr{F}$, then $\bigcap_{i=1}^{\infty} A_i \in \mathscr{F}$.

1.12. THE CONCEPT OF A THEORY

The comments on the notion of a theory to be found in this section are preliminary. We shall examine this concept in a more detailed manner in Chapter III.

By a *theory in the deductive sense* (a *deductive theory*) we shall mean any deductively closed set of sentences. If Θ is a theory, the language in which Θ is stated will usually be denoted as $\mathscr{L}_{\Theta}$.

Elements of a theory will be referred to as its *theorems*. In the case of empirical theories we shall occasionally use also the term "law". The sentences of the language of a theory will often be referred to as *sentences* of the theory, or *hypotheses*.

When empirical theories are considered it is often convenient to view a theory as a structure which apart from a theory in the deductive sense and its language involves also some additional constituents, for instance the set of all empirical systems to which the theory refers, or the set of all testing procedures permitted in the theory. In such a case the set of theorems (i.e. the deductive theory) will often be referred to as the *content of the theory* in question.

Let me mention a few methodological concepts applied in metatheoretical investigations.

A theory Θ is said to be *deductively complete* if and only if it is consistent and for all α in $\mathscr{L}_\Theta$ either α or $\neg\alpha$ is a theorem of Θ.

Let $\mathscr{K} \subseteq Int(\mathscr{L}_\Theta)$. The theory Θ will be said to be *$\mathscr{K}$-complete* if and only if the intersection

(1) $\bigcap\{Val(I): I \in \mathscr{K}\}$

is a subset of Θ.

Note that

ASSERTION 1.12.1. *For each $\mathscr{K} \subseteq Int(\mathscr{L})$, the intersection* (1) *is deductively closed, thus is a deductive theory.*

In what follows the theory defined as the intersection (1) will be denoted as $Th(\mathscr{K})$.

The following facts are of a certain importance:

ASSERTION 1.12.2. *For each set of sentences X,*

$$Mod(Th(Mod(X)) = Mod(X).$$

ASSERTION 1.12.3. *For each set of standard interpretations $\mathscr{K}$,*

$$Th(Mod(\mathscr{K})) = Th(\mathscr{K}).$$

The proofs of the two assertions are trivial. Note also that

ASSERTION 1.12.4. *For each standard interpretation I, Th(I) is a deductively complete theory.*

II. REGULARITIES

2.0. TWO CONVENTIONS

CONVENTION 2.0.1. **Throughout this chapter, if not explicitly stated otherwise, by a system we shall mean an elementary system.**

CONVENTION 2.0.2. Whenever a symbol of the form $\mathfrak{a}(t)$ or $\mathfrak{a}_{/u}$ used, it will be taken for granted that $t \in T(\mathfrak{a})$ or respectively $u \subseteq T(\mathfrak{a})$.

2.1. TWO TYPES OF REGULARITIES

When speaking of a regularity which governs a phenomenon $\mathscr{Y}$ we shall always mean either some property common to all states of $\mathscr{Y}$ or some relation which defines the way in which states of $\mathscr{Y}$ transform in time. In both cases a regularity can be defined to be a property of a phenomenon, and therefore in what follows instead of saying that "a phenomenon $\mathscr{P}$ displays regularity **R**" or "a phenomenon $\mathscr{P}$ falls under **R**" we shall sometimes write $\mathscr{P} \in \mathbf{R}$.

The regularities of the first of the two kinds we have mentioned will be referred to as *structural*, those of the second kind will be called *dynamic*. These explanations do not claim to be formal definitions. It is quite probable that such definitions can be given, but the task would be rather unrewarding. For the sake of our discussion it is enough to characterize these notions by means of suitably chosen examples. This we shall do in this section. Subsequent sections will be devoted to a more detailed analysis of some special kinds of regularities of the two types considered.

EXAMPLE 1. Let $\mathscr{Y}$ be an empirical phenomenon with the following properties:

(i) The set of all states $St(\mathscr{Y})$ of $\mathscr{Y}$ is finite. Suppose for instance that:

(1) $$St(\mathscr{Y}) = \{s_1, s_2, s_3\}.$$

(ii) Each system $\mathfrak{a} \in \mathscr{X}$ is in each of the states $\mathbf{x}_1, \mathbf{x}_2, \mathbf{x}_3$ at most once in its whole history,

(iii) $\mathbf{x}_1$ never transforms directly into $\mathbf{x}_3$, i.e. for each $\mathfrak{a}$ in $\mathscr{X}$ and for each t_1, t_2 such that $t_1 < t_2$, whenever $\mathfrak{a}(t_1) = \mathbf{x}_1, \mathfrak{a}(t_2) = \mathbf{x}_3$, then for some $t \in [t_1, t_2]$, $\mathfrak{a}(t) = \mathbf{x}_2$.

If the fact expressed by the equality (1) does not hold by the conceptualization of $\mathscr{X}$, in other words, if the set of possible states $St^*(\mathscr{X})$ is greater than $St(\mathscr{X})$, then that fact has an empirical character and constitutes some regularity of the structural type. The law of quantum mechanics stating that in a stationary condition there can only be definite energies for an atom describes the regularity being an exemplification of (i).

Both Clause (ii) and (iii) concern the way of transformation of states of $\mathscr{X}$ and hence they both characterize certain dynamic regularities. Transformation of a larva into a chrysalis and then into a butterfly may serve as an illustration of those two regularities.

EXAMPLE 2. Consider a phenomenon $\mathscr{X}$ (which need not be the same as discussed earlier) and assume that for all $\mathfrak{a}$ in $\mathscr{X}$ and for all t,

(2) $\mathfrak{a}(t) = \mathfrak{a}(t+r)$,

provided that t and $t+r$ are in $T(\mathfrak{a})$ (Incidentally, in view of Convention 2.0.2 this qualification is automatically valid and thus need not be stated explicitly).

Evidently the relation described in Example 2 is a regularity. Equality (2), although does not define univocally the way in which states of $\mathscr{X}$ undergo transformation, provides us with much information about it. States of systems in $\mathscr{P}$ undergo periodic transformations and, moreover, the period of repetition of states of the systems is common to all of them.

While discussing Example 2 it seems advisable to describe a situation when all intuitions connected with the notion of regularity begin to fail and we are not sure whether a given theorem, quite general in character, defines any regularity or not. Let $\mathbf{x}_0 \in St^*(\mathscr{X})$. As an immediate consequence of (2) we have

(3) If $\mathfrak{a}(t) = \mathbf{x}_0$, then $\mathfrak{a}(t+r) = \mathbf{x}_0$.

If $\mathbf{x}_0$ is a state which "often" occurs in the course of $\mathscr{X}$, eg. can be observed in almost all cases of $\mathscr{X}$, then the relation (3) is very likely to be an unquestionable regularity. Suppose, however, that the state $\mathbf{x}_0$ is never realized, i.e. $\mathfrak{a}(t) \neq \mathbf{x}_0$ for all $\mathfrak{a} \in \mathscr{X}$ and for all $t \in T(\mathfrak{a})$. Clearly, implication (3) remains still valid because it is vacuously valid. Nevertheless, to call the relation expressed by this implication a regularity does not seem to be justified.

The problem should not be considered very serious. If regularities may be "stronger" or "weaker", for instance in the sense that they rule out some greater or smaller number of possible histories as those that may not be actually carried into effect, we may assume that in the extremal case a regularity may be of "vacuous" character. If we decide to admit the existence of vacuous regularities, then each general hypothesis, i.e. the one concerning some phenomenon $\mathscr{X}$ taken as a whole (e.g. concerning all states or all histories of $\mathscr{X}$ is a certain regularity, provided that it gives an adequate account of the phenomenon. A word of explanation is needed here. The impression which could arise that we have reduced the notion of regularity to the notion of law is erroneous. Regardless of how rich is the formal apparatus we have at our disposal there may exist regularities which cannot be described by means of the notions and formal devices we are able to use.

EXAMPLE 3. Suppose that $\mathscr{X}$ is composed of systems which involve variables F_1, F_2, F_3. Assume also that the variables can take arbitrary values, but the inequalities:

$$0 \leqslant F_1 + F_2 + F_3 \leqslant 10 \tag{4}$$

are always satisfied (more precisely: $0 \leqslant F_1(t) + F_2(t) + F_3(t) \leqslant 10$).

The relation (4) excludes occurrence of some of the possible states of the phenomenon $\mathscr{X}$ and therefore it undoubtedly has the character of a structural regularity. The next example appears to be slightly more baffling.

EXAMPLE 4. Let $\mathscr{Y}$ be like in Example 3. Assume that in all cases of the phenomenon $\mathscr{Y}$ the equality

(5) $$F_3 = \frac{dF_1}{dt} + \frac{dF_2}{dt}$$

holds.

Apparently regularities (4) and (5) seem to be of the same kind. But already after a moment of thought one sees that equation (5) need not (though it can) exclude occurrence of any of the possible states of $\mathscr{Y}$. It excludes for sure only some possible histories of the instances of $\mathscr{Y}$. Following rigorously our distinction we ought to consider (5) a dynamic regularity keeping, however, in mind that this regularity can be combined in nature, i.e. it can be both dynamic and structural. Thus the example just discussed shows that the distinction between the two types of regularities accepted by us cannot be carried out in a consequent and unique way.

An evident similarity of situations in Examples 3 and 4 induces us to give some common name to the regularities spoken about in these examples. Since the regularities in question concern relations which system variables bear among themselves, we shall call them *correlational.* As before, the explanations given should not be treated as a definition. In one of subsequent sections we shall discuss correlational regularities in a more detailed way.

The introduced distinctions meet the division of regularities into *stochastic* and *non-stochastic.* All the examples given so far are the examples of non-stochastic regularities, i.e. the regularities which can be defined without any reference to the notion of probability. Such regularities are sometimes called *deterministic.* However, in the sequel the notion of determinism will be burdened with some special sense, stochastic determinism being not excluded.

EXAMPLE 5. Let $\mathscr{Y}$ be like in Example 1. Assume that the probability of a randomly chosen system a to be at a randomly chosen time instant t at one of the states $\mathbf{x}_1, \mathbf{x}_2, \mathbf{x}_3$ equals 0.5, 0.2, 0.3, respectively, in symbols:

(6) $$P(x_1) = 0.5, \quad P(x_2) = 0.2, \quad P(x_3) = 0.3.$$

Evidently that is a stochastic counterpart of the regularity defined by Clause (i) in Example 1. A more detailed analysis of this example is left to the reader. By a simple argument referring to properties of the notion of probability one can show that the regularity defined in Example 5 is not dynamic. The reason is that knowing a state of a system $\mathfrak{a} \in \mathscr{L}$ at some time instant t and using only equalities (6) one cannot draw any conclusions concerning preceding or subsequent states of the system.

EXAMPLE 6. Assume, like in Example 1, that $St(\mathscr{L}) = \{\mathbf{x}_1, \mathbf{x}_2, \mathbf{x}_3\}$, revoking, however, assumptions (ii) and (iii). Suppose that we watch the course of the phenomenon $\mathscr{L}$ observing various instances of this phenomenon at some regular intervals r. In other words, the subject of our observations are sequences of the form

(7) $$\mathfrak{a}(t), \mathfrak{a}(t+r), \ldots, \mathfrak{a}(t+kr), \ldots$$

where $\mathfrak{a} \in \mathscr{L}$ and t is an arbitrary instant of time from $T(\mathfrak{a})$. Assume further that in the ways of transformation of the states of $\mathscr{L}$ in sequences (7) no non-stochastic regularity can be found (that, of course, does not mean that no regularity of this type exists) but on the other hand the following regularity has been discovered. If in a sequence of the form (7) chosen in a random way (i.e. we choose randomly both the system $\mathfrak{a}$ and the initial time instant t) at some moment $t+kr$ a state $\mathbf{x}_i$ has been realized, then with the probability p_{ij} the next state, i.e. the state at the instant $t+(k+1)r$, will be x_j. In symbols this can be written as follows:

(8) $$P(\mathbf{x}_i \rightarrow \mathbf{x}_j) = p_{ij}.$$

In the present example as well as in the previous one, when speaking of probability we assumed tacitly the objective and at the same time frequentistic interpretation of probability. The former assumption seems to be necessary if equalities (6) and (8) are to express some property of the phenomenon $\mathscr{L}$, the latter one provides the term probability with a definite meaning suitable for the contexts in which it has been used. Clearly there is a great deal of functions which satisfy axioms of probability calculus and at the same time take values dependent on the nature of empirical events on which they have been defined. There is no sensible reason, however, to assume that the probability spoken about can be any such a function whatsoever.

The reader can find a large number of textbooks and studies devoted to the notion of objective probability and to the ways of handling it in actual empirical situations. Therefore it does not seem advisable to analyse the notion in the present work. One point, however, deserves our special attention.

Suppose that $\mathscr{X}$ is a definite phenomenon and the states $\mathbf{x}_1, \mathbf{x}_2, \mathbf{x}_3$ have been precisely defined. Assume also, as it has been done in the examples discussed, that the states do not transform in a unique way; in sequences of the form (7) the state $\mathbf{x}_i$ is not always followed immediately by the same state $\mathbf{x}_j$. In such a situation we are inclined to search for a stochastic regularity in transformation of the states of the system under investigation.

It may turn out that we are unable to discover such a regularity. Experiments carried out according to a well known procedure can lead to diverging results. Obviously, we may be mistaken when stating that probabilities of transformations $\mathbf{x}_i \rightarrow \mathbf{x}_j$ are undefinable. It could happen that we had no patience enough to continue experiments up to positive results, or perhaps our experiments were carried out in some incorrect way, for example the applied methods of random choice of the sequences (7) were, in fact, not random and thus in various experiments various groups of instances of $\mathscr{X}$ were distinguished. We may happen to be wrong in our belief that the probability of the transformations considered is undefinable. On the other hand, when we use the notion of probability in an empirical sense we have to be prepared for the possibility that no function which both satisfies the axioms of probability calculus and takes the values which can be determined by means of accepted empirical techniques, exists (cf. S. Nowak, 1972, for a discussion of the caution which has to be preserved when using the concept of probability).

In our further considerations the problems of stochastic regularities will only occasionally be touched upon. We shall focus our attention on non-stochastic regularities. It is rather clear that each regularity of non-stochastic type has its stochastic counterpart. It is also clear that in scientific researches, because of various well-known reasons (for example a great complexity of a phenomenon due to a great number of variables on which the phenomenon depends), we have to put up with stochastic regularities.

The problems connected with stochastic regularities are certainly of great importance. However, if we decided to analyse them in a more detailed way so that the discussion would not be a repetition, only the terminology being suitably modified, of what has already been said about non-stochastic regularities, we would have to apply a greater formal apparatus than we really wish to use here and to enlarge to a great extent the considerations which we want to present.

2.2. STATE-DETERMINED PHENOMENA

A phenomenon $\mathscr{X}$ is called state-determined if its states transform in time in a unique way. The essential sense of this explanation is best understood if the course of a phenomenon is observed with respect to some sequence of instants $\{t_n\}$.

DEFINITION 2.2.1. *A phenomenon* $\mathscr{X}$ *is* state-determined with respect to $\{t_n\}$ If and only if *there exists a one-to-one function*

$$D: St(\mathscr{X}) \rightarrow St(\mathscr{X})$$

such that for any $\mathfrak{a} \in \mathscr{X}$, *and any* t_i *in* $\{t_n\}$,

$$\mathfrak{a}(t_{i+1}) = D(\mathfrak{a}(t_i)).$$

(Observe that according to Convention 2.0.2. t_i and t_{i+1} should be chosen so as to belong to $T(\mathfrak{a})$!)

Given an arbitrary initial state $\mathbf{x}_0$ and using the function D we may determine any of the states that follow or precede $\mathbf{x}_0$ in exactly the same way, regardless of both the system on which the state $\mathbf{x}_0$ was realized and the moment of occurrence of $\mathbf{x}_0$. The states form the following sequence:

$$(1) \qquad D^{-1}(D^{-1}(\mathbf{x}_0)), D^{-1}(\mathbf{x}_0), \quad \mathbf{x}_0, D(\mathbf{x}_0), D(D(\mathbf{x}_0)), \ldots$$

The function D is called a *transition function.* In the definition given above it is assumed to be one-to-one. This means that we require both the earlier states to determine uniquely the later ones (this type of determinism is sometimes called *prospective* or *futuristic*) and the later states to determine the earlier ones (*retrospective* or *teleological* determinism). The determinism which is both prospective and retrospective is sometimes called reversible. This is exactly this sort of determinism that we have defined.

Any dynamic regularity possesses its prospective, retrospective and reversible variants. The thing is so obvious, however, that we see no sense in defining all three variants each time when a new regularity is discussed. Usually we shall confine ourselves to the reversible variant only.

The definition of a phenomenon state-determined with respect to discrete time, i.e. represented by a sequence of time instants, can be in an obvious way generalized onto the continuous case. We shall do this by means of the following definition.

DEFINITION 2.2.2. *A phenomenon* $\mathscr{X}$ *is* state-determined *if and only if for all* $\mathfrak{a}$, $\mathfrak{b}$ *in* $\mathscr{X}$, *for all instants* t, s *if*

(i) $\mathfrak{a}(t) = \mathfrak{b}(s)$,

then for all r,

(ii) $\mathfrak{a}(t+r) = \mathfrak{b}(s+r)$.

Observe that each state-determined phenomenon $\mathscr{X}$ has the following characteristic property. The lines of behaviour (cf. Section 1.7.) of the systems of $\mathscr{X}$ form in $(n+1)$-Euclidean space (where n is the number of variables the phenomenon involves) a set of curves such that through no point of the space more than one such curve runs. Indeed if two lines of behaviour coincide at any pair of points (and thus clause (i) is satisfied), then by (ii) they must coincide at any point at which they are defined.

The following alternative characteristic of state-determined phenomena is an obvious consequence of the definition.

COROLLARY 2.2.3. A phenomenon $\mathscr{X}$ is state-determined if and only if there exists a function $D_{\mathscr{X}}: St(\mathscr{X}) \times Re \rightarrow St(\mathscr{X})$ such that for each $\mathfrak{a}$ in $\mathscr{X}$ and for each t,

$$D_{\mathscr{X}}(\mathfrak{a}(t), r) = \mathfrak{a}(t+r).$$

The function $D_{\mathscr{X}}$ is a counterpart of the transition function D defined earlier and plays a similar role. In what follows we shall call $D_{\mathscr{X}}$ the *rule of state transformations* for $\mathscr{X}$, or shortly the *state-transformation*.

Connection between discrete and continuous case is established by the following, obvious assertion.

ASSERTION 2.2.4. *$\mathscr{X}$ is state-determined if and only if $\mathscr{X}$ is state--determined with respect to all time sequences $\{t_n\}$ such that the interval $[t_i, t_{i-1}]$ is constant for all i.*

One may also easily prove that:

ASSERTION 2.2.5. *The following conditions are equivalent:*

(i) *$\mathscr{X}$ is state-determined,*

(ii) *all subsets $\mathscr{X}' \subseteq \mathscr{X}$ are state-determined,*

(iii) *all phenomena $\mathscr{X}'$ such that the history of each system $\mathfrak{a}'$ in $\mathscr{X}'$ coincides up to a translation in time with a subhistory of an $\mathfrak{a}$ in $\mathscr{X}$ are determined.*

An important thing about the rather trivial fact announced by Assertion 2.2.5. is that an analogon of this assertion is valid for all the regularities we shall discuss. Each of them may be conceived as a property of phenomena shared by all subphenomena of any phenomenon which possesses it. We shall say that $\mathscr{X}'$ is a *subphenomenon* of $\mathscr{X}$ if and only if systems in $\mathscr{X}'$ are time restrictions (proper or not) of systems in $\mathscr{X}$. In particular when $\mathscr{X}$ falls under a certain regularity **R** of this sort each singular phenomenon $\mathfrak{a}$, where $\mathfrak{a}$ is in $\mathscr{X}$, falls under it as well (in the sense that $\{\mathfrak{a}\}$ does).

Observe that if a regularity of $\mathscr{X}$ is inherited by singular phenomena corresponding to the instances of $\mathscr{X}$, it need not, in general, be the other way round. For example, if all $\mathfrak{a}$ in $\mathscr{X}$ are state-determined, it does not mean that $\mathscr{X}$ must be state--determined as well. The rules of state transformations of the particular systems in $\mathscr{X}$ may differ among themselves and there need not exist any state transformation adequate for $\mathscr{X}$ as a whole.

The fact that the regularities which we are going to discuss are inherited by subphenomena is due to our decision to restrict our considerations to non-stochastic regularities. It is rather clear that when **R** is a stochastic property of $\mathscr{Y}$ one should not expect that **R** is shared by all subsets of $\mathscr{Y}$ or, more generaly, by all subphenomena of $\mathscr{Y}$.

Let us pass on to another subject. Although in order to simplify the matter we have restricted our considerations to elementary systems, this can perhaps be the most convenient moment to explain how a more general approach is possible. Besides, it will turn out that the results of our discussion concerning generic and partially also complex systems are relevant in an important way to elementary systems.

Let then $\mathscr{P}$ be a phenomenon which need not consist of elementary systems. The way in which the notion of state-determinism should be defined so that it could be applicable to $\mathscr{P}$, regardless of whether it contains generic systems or not, seems to be obvious.

DEFINITION 2.2.6. *A phenomenon* $\mathscr{P}$ *(which may consist of generic systems) is* state-determined *if and only if for all* $\mathfrak{A}$, $\mathfrak{B}$ *in* $\mathscr{P}$ *for all* t, s *and for all mappings* $f: A \to B$, *if*

(i) $\mathfrak{A}(t) =_f \mathfrak{B}(s)$,

then for all $r \geqslant 0$

(ii) $\mathfrak{A}(t+r) =_f \mathfrak{B}(s+r)$.

Definition 2.2.2. is easily seen to be a special case of Definition 2.2.6., and thus the problem we posed possesses a simple and apparently obvious solution. There is, however, a fly in the ointment.

The definition proposed by no means restricts the way in which the mapping f is to be selected. It may happen, however, that the sets A and B have a certain "intrinsic structure" which is not determined merely by the states of those systems. Imagine, for instance, that A and B are two groups of animals, say two packs of wolves, which at a given moment of time are in the same state up to a certain mapping f (with respect to variables in terms of which they are examined). If f assigns the wolf that dominates all members of the pack A to the wolf that is submitted to all members of B, most likely the way of comparing the systems $\mathfrak{A}$ and $\mathfrak{B}$ is misleading.

Let us consider another more rigorous example. Put $A = \{a_1, a_2, a_3\}$, and $B = \{b_1, b_2, b_3\}$ and let a_i, b_i $(i = 1, 2, 3)$ be mechanical particles. Assume that a_1, a_2, and similarly b_1, b_2, are constrained by a rigid non-deformable coupling. No constraints are put on either a_3 or b_3. Assuming that the systems $\mathfrak{A}$, $\mathfrak{B}$ are examined in terms of standard variables of particle mechanics, we can easily express the fact of existence of constraints in terms of variables of the systems. The thing amounts to the fact that the distance between a_1 and a_2 as well as the distance between b_1 and b_2 are constant. Observe, however, that the existence of constraints cannot be discovered merely by comparing two singular states of the systems, regardless of which states have been picked out to be compared. We assume here that a qualitative parameter "are constrained by a rigid non-deformable coupling" has not been included in the conceptualization of the systems.

In the situation described it may happen that $\mathfrak{A}(t) =_f \mathfrak{B}(s)$ under a mapping f which is defined in such a way that a constrained particle of one system is assigned to a free one of the other, i.e. the condition:

(2) $\qquad f(a_3) = b_3$

is not satisfied. But, for obvious reasons, such an f does not provide a proper way of comparing $\mathfrak{A}$ and $\mathfrak{B}$, and the fact that $\mathfrak{A}(t)$ coincides up to f with $\mathfrak{B}(s)$ does not suffice by itself to draw the conclusion that the future states of the two systems will coincide as well, although we may reasonably assume that both $\mathfrak{A}$ and $\mathfrak{B}$ are state-determined.

The simplest way to cope with the problem on its formal level is to relativise the notion of state-determinism to a group Γ of mappings (cf. R. Wójcicki, 1975). Then a phenomenon $\mathscr{P}$ would be said to be *state-determined with respect to* Γ when the definiens of 2.2.6. holds true with the proviso that f is in Γ. Clearly, dealing with a specific phenomenon $\mathscr{P}$ one should select Γ so that it conforms to the intrinsic structure of systems in $\mathscr{P}$. What is to be meant by the intrinsic structure of $\mathscr{P}$ depends on the nature of $\mathscr{P}$. Observe however that instead of formulating the problem in a rather obscure way: "discover the intrinsic structure of systems in $\mathscr{P}$", we may formulate it quite precisely: "find such a group of transformations Γ that $\mathscr{P}$ is state-determined under Γ".

Perhaps, in connection with the last remark I owe the reader some additional explanation. Although in general the notion of intrinsic structure has no clear conotation, none the less it can, in certain special cases, be defined in a rigorous way. Note for instance that the intrinsic structure of complex systems can be identified with the network of couplings by means of which the systems being parts of some complex system are joined among themselves. Whenever we compare states of two complex systems $\mathfrak{R}$, $\mathfrak{R}'$, say by comparing the elementary systems $\mathfrak{a}_{\mathfrak{R}}$, $\mathfrak{a}_{\mathfrak{R}'}$ by means of which they are represented (cf. Section 1.5), we must not ignore the networks of couplings of the two systems, otherwise the results of the comparisons will be misleading. Incidentally, a great deal of systems investigated in science are complex what, at least to some extent, proves that the problem of intrinsic structure of phenomena is of considerable significance.

What can be the counterpart of this problem when elementary systems are considered? Since some of generic as well as some of complex systems can be represented by elementary systems it is clear that the problem cannot simply disappear. Let us then examine the matter closer. For that purpose let us turn back to the two systems $\mathfrak{A}$, $\mathfrak{B}$ of particles we have described above.

As we remember (cf. Section 1.3), all finite generic systems can be uniquely represented by elementary systems. In particular, the systems $\mathfrak{A}$ and $\mathfrak{B}$ can be uniquely represented by certain elementary systems $\mathfrak{a}_{\mathfrak{A}}$ and $\mathfrak{a}_{\mathfrak{B}}$. The problem of restricting mappings from A into B so that the mappings which fail to satisfy Condition (2) are excluded does not disappear. It simply becomes that of ordering variables of the systems $\mathfrak{a}_{\mathfrak{A}}$ and $\mathfrak{a}_{\mathfrak{B}}$ in a proper way. Each of the variables refers now to one particle only (thus the position of a_1 and that of a_2 constitute two different parameters) and when states of systems $\mathfrak{A}$ and $\mathfrak{B}$ are being compared we have to be sure that the variables in $\mathfrak{A}$ that refer to a_3 are compared with those that, in $\mathfrak{B}$, refer to b_3. The standard way of defining $\mathfrak{A}$ and $\mathfrak{B}$ to be elementary systems involves the notion of configurational space and is more sophisticated than our simple-minded approach, but with respect to what we are interested in that difference is irrelevant.

2.3. MATHEMATICAL MODELS FOR STATE TRANSFORMATIONS

Let $\mathscr{X}$ be a state-determined phenomenon, and let $D_{\mathscr{X}}$ be the rule of transformations of states of $\mathscr{X}$. Observe that $D_{\mathscr{X}}: Re^{n+1} \to Re^{n}$, where n is the number of system variables of the conceptualization of $\mathscr{X}$, and thus is a function which, in principle, can be defined in a purely formal way, provided that the mathematical apparatus available is not too weak. In fact, as it is known, no matter how strong the mathematical formalism we have at our disposal is, it is never strong enough to allow us to define all functions of the kind of the function $D_{\mathscr{X}}$. This is simply a matter of an arithmetic of cardinal numbers: for each $n \geqslant 1$ there is more than continuum of functions that map Re^{n+1} into Re^{n}, at the same time for any language whose vocabulary is recursively definable and whose sentential formulas are strings of elementary symbols, the cardinal number of equations which can be stated in that language must be countable.

In view of these remarks the question of whether $D_{\mathscr{X}}$ can be modelled by mathematical means or not must therefore be relativised to a given mathematical theory or equivalently to the mathematical formalism of a given empirical theory.

A set of equations by means of which the state transformation $D_{\mathscr{X}}$ can be modelled in the language of an empirical theory (provided that it can be modelled at all) can take various forms. It is then of considerable importance to have a clear idea what is to be meant by a (mathematical) model of $D_{\mathscr{X}}$. The concept of a mathematical model will be discussed in a more detailed way in the next chapter, here we shall confine ourselves to some preliminary remarks.

Let us examine the following answer to the question posed.

A set of equations E is a *mathematical model* of $D_{\mathscr{X}}$ if and only if the following two conditions are satisfied:

(+) The set of all realizations (semantic models) for E is state--determined.

(+ +) Each system a in $\mathscr{X}$ is a realization of E.

Denote the set of realizations for E by $Mod(E)$. If $Mod(E)$ is state-determined, then (by Assertion 2.2.3) there exists a state transformation $D_{Mod(E)}$ of the set $Mod(E)$. Since, by (+ +) $\mathscr{X} \subseteq Mod(E)$, then $D_{\mathscr{X}}$ must be the restriction of $D_{Mod(E)}$ to $St(\mathscr{X})$ and this motivates the answer proposed.

The presented argument need not be, however, fully convincing. The trouble is that as a rule, equations are given in a form of open formulas, whereas the notion of a realization (Definition 1.9.2) has been defined to be applicable to set of sentences only. We have then to examine this point.

Indeed, the equations which are to characterize $D_{\mathscr{X}}$ can, for example, take the following form

(1) $$d(x_1, \ldots, x_n, r) = (y_1, \ldots, y_n),$$

or equivalently (in a notation which results from replacing vectors by its components) the form

(2) $$d_i(x_1, \ldots, x_n, r) = y_i, \quad i = 1, \ldots, n.$$

The symbol d (or its components d_i) should not be treated as a symbol of a function. It can be a complicated expression involving a number of mathematical operations. None the less the idea which underlies the equation (1) is that $d(x_1, \ldots, x_n, r)$ conforms to $D_{\mathscr{X}}(x_1, \ldots, x_n, r)$ in the following sense:

(§) For all x_i, y_i, $i = 1, \ldots, n$, for all a in $\mathscr{X}$, and for all t_0 and r, the following is satisfied: If (1) and $a(t_0) = (x_1, \ldots, x_n)$, then $a(t_0 + r) = (y_1, \ldots, y_n)$.

Obviously the condition (§) is equivalent to the following

(§§) For all a in $\mathscr{X}$, for all t_0, and for all r:

$$d_i(F_1^a(t_0), \ldots, F_n^a(t_n), r) = F_i^a(t_0 + r),$$

$i = 1, \ldots, n$.

(The superscript a indicates that the F_i's are measured with respect to a).

In turn (§§) can be reformulated as follows:

(§§§) For all a in $\mathscr{X}$, a is a realization of the following sentence:

$$\forall t_0 \, \forall r \, d_1(F_1(t_0), \ldots, F_n(t_0), r) = F_1(t_0 + r) \wedge \ldots \wedge d_n(F_n(t_0), \ldots, F_n(t_0), r) = F_n(t_0 + r).$$

The components of the conjunction stated in the parentheses are the equations (2) written in the form which brings to light the very nature of the symbols $x_1, \ldots, x_n$, $y_1, \ldots, y_n$ and their dependence on time. In this way condition (§§§) makes it intelligible how equations might correspond to certain sentences and in fact serve as an abbreviated and thus more convenient version of the latter. There should be no obstacle then to treat equations as sentences. This removes the doubt that was raised. Note also that (§§§) specifies conditions (+) and (+ +) and thus the argument by which we arrived at (§§§) can be treated as a justification of those conditions.

Incidentally, the most typical form of the equations which are to characterize the state transformation of a state-determined phenomenon is the following:

$$(3) \qquad \frac{dx_i}{dt} = f_i(x_1, \ldots, x_n), \qquad i = 1, \ldots, n.$$

Thus the equations do not involve time explicitely, i.e. in the form of a separate parameter such as r in (2), and determine the first derivative of the variables F_i rather than F_i's directly.

It can be proved (for details cf. W. R. Ashby, 1960) that whenever equations (2) exist, they can be replaced by equations of the form (3). These are technicalities, however, which are only loosely related the subject of our discussion.

2.4. HISTORY-DETERMINED PHENOMENA

The histories of instances of a phenomenon need not be determined by their present states but still the phenomenon may be deterministic in the following sense.

DEFINITION 2.4.1. *A phenomenon* $\mathscr{X}$ *is* past-determined (future determined) *if and only if for all* $\mathfrak{a}$, $\mathfrak{b}$ *in* $\mathscr{X}$, *for all instants* t_1, t_2; *if for all* $r \geqslant 0$, $(r \leqslant 0)$,

(i) $\mathfrak{a}/[t_1 - r, t_1] = \mathfrak{b}/[t_2 - r, t_2]$,

then for all $s \geqslant 0$, $(s \leqslant 0)$,

(ii) $\mathfrak{a}(t_1 + s) = \mathfrak{b}(t_2 + s)$.

A phenomenon will be called *history-determined* if it is both past and future determined. Clearly we immediately have:

COROLLARY 2.4.2. *Each state-determined phenomenon is history-determined.*

The converse, of course, need not be true. The difference between inborn and learned behaviour provides an exemplification of the difference between the state-determined and history-determined phenomena. The scenaris of animals season migrations, say storks from Europe to Africa, is within a species the same. Similarly mating ritual of, for instance, elephants or pigeons. These are state-determined phenomena. In turn, to explain why two different rats behave during an experiment in maze-running in a different way we must take into account the past-experience of the animals. The phenomenon (when defined in a suitable manner) is past determined, being not state-determined.

The type of determinism we are discussing now involves in a natural way the notion of *memory*. If a system behaves in a history-determined way it behaves as if it remembered its own history. But it need not remember or perhaps "remember" the whole history. Its memory may be limited in time.

DEFINITION 2.4.3.

A. *A phenomenon* $\mathscr{X}$ *is* determined by its past (future) histories of the length r $(r \geqslant 0)$ *if and only if the following condition is satisfied*: *for all* $\mathfrak{a}$, $\mathfrak{b}$ *in* $\mathscr{X}$, *for all instants* t_1, t_2, *if*

(i) $\mathfrak{a}/[t_1 - r', t_1] = \mathfrak{b}/[t_2 - r', t_2]$

for some $r' > r$ $(-r > r)$, *then for all* $s \geqslant 0$ $(s \leqslant 0)$,

(ii) $\mathfrak{a}(t_1 + s) = \mathfrak{b}(t_2 + s)$.

B. *If $\mathscr{Z}$ is determined both by past and future histories of the length* 0 *we shall say that it is* almost state-determined.

The number $r_{\mathscr{Z}}$ being the least of all numbers r such that $\mathscr{Z}$ is determined by past histories of the length r is what may be called the *length of the memory of* $\mathscr{Z}$. Obviously for such a number to exist $\mathscr{Z}$ must be past-determined and what is more, it must be determined by its past histories of finite length. If the latter is not true although the former is, the phenomenon $\mathscr{Z}$ may be said to have an *unlimited memory*. Observe that the fact that memory of $\mathscr{Z}$ is unlimited need not imply that the memories of all instances of $\mathscr{Z}$ are unlimited as well. It may happen that all of them have finite memory in the sense that each singular phenomenon $\mathfrak{a}$ is determined by past histories of some finite length $r_{\mathfrak{a}}$, but nevertheless the memory of $\mathscr{Z}$ as a whole is not finite. Clearly such a situation takes place when the set of all $r_{\mathfrak{a}}$'s has no upper bound.

It should be obvious that the term "memory" applied in the way we have done here need not refer to the memory treated as a psychological phenomenon. Observe for instance that the following assertion is valid.

ASSERTION 2.4.5. *If $\mathscr{Z}$ is a* periodic phenomenon *in the sense that there exists a least number $r_{\mathscr{Z}}$ such that for all $\mathfrak{a}$ in $\mathscr{Z}$, and for all t,*

$$\mathfrak{a}(t) = \mathfrak{a}(t+r),$$

then Z is determined by past histories of the length $r \geqslant r_{\mathscr{Z}}$.

A proof of the assertion can be established by an easy argument. It is obvious that phenomena of any kind whatsoever, say mechanical, may be periodic and thus furnished with a "memory" in the sense discussed here.

Curiously enough phenomena with the length of memory 0, i.e. almost state-determined, need not be state-determined. For an example consider a mechanical particle a whose state is characterized by position s, force F, and mass m, so consider the system

$$\mathfrak{a} = (a, s, F, m).$$

Given an initial state $\mathfrak{a}(t_0)$ of $\mathfrak{a}$, what we need in order to calculate any future state $\mathfrak{a}(t)$ is the value of the velocity $\mathbf{v} = \frac{d}{dt} s$ at time t_0. Now to calculate $\mathbf{v}(t_0)$ it is enough to know the values of s within any finite interval $[t_0 - r, t_0]$, $r > 0$. Obviously, when the history of $\mathfrak{a}$ within such an interval is known, in particular the values of s are established within it. Since r can be arbitrarily small, then under the assumption that the behaviour of $\mathfrak{a}$ is governed by the laws of particle mechanics, the system $\mathfrak{a}$ (or equivalently the singular phenomenon $\mathfrak{a}$) is almost state-determined but is not state-determined.

We have, however, the following

THEOREM 2.4.6. *Let $\mathscr{X}$ be a phenomenon whose elements are systems of the form $(a, F_1, \dots, F_n)$, and let for each $\mathfrak{a} \in \mathscr{X}$ the variables $F_i^{\mathfrak{a}}$ be differentiable with respect to time. Denote by $\mathscr{X}'$ the phenomenon which results from $\mathscr{X}$ by expanding the systems in $\mathscr{X}$ to systems of the form $(a, F_1, \dots, F_n, dF_1/dt, \dots, dF_n/dt)$. Under the assumptions stated above $\mathscr{X}$ is almost state-determined if and only if $\mathscr{X}'$ is state-determined.*

PROOF. Suppose first that $\mathscr{X}'$ is state-determined, and assume that $\mathfrak{a}/[t-r, t] = \mathfrak{b}/[s-r, s]$ for some $\mathfrak{a}, \mathfrak{b} \in \mathscr{X}$ and for some t, s and r. Then clearly $\mathfrak{a}'(t) = \mathfrak{b}'(s)$, where $\mathfrak{a}', \mathfrak{b}'$, are expansions of $\mathfrak{a}$ and $\mathfrak{b}$, respectively, in $\mathscr{X}$. Consequently $\mathfrak{a}'(t+r') = \mathfrak{b}'(s+r')$ and therefore also $\mathfrak{a}(t+r') = \mathfrak{b}(s+r')$ for all r'. The r may be selected to be arbitrarily small, hence $\mathscr{X}$ is determined by histories of length 0 and thus almost state-determined.

In turn, suppose that $\mathscr{X}'$ is not state-determined. Then, for some $\mathfrak{a}', \mathfrak{b}'$ in $\mathscr{X}'$ and for some t, s and r, $\mathfrak{a}(t) = \mathfrak{b}(s)$ but $\mathfrak{a}(t+r) \neq \mathfrak{b}(s+r)$. Let r_0 be the least real number within the interval $[0, r]$ such that

(1) $$\mathfrak{a}(t+r_0) \neq \mathfrak{b}(s+r_0).$$

The existence of such a number in guaranteed by the fact that the set of all real numbers is well-ordered. Consider the intervals $[t, t+r_1]$, $[s, s+r_1]$, where r_1 is any real number between 0 and r_0. We clearly have

$$\mathfrak{a}'/[t, t+r_1] = \mathfrak{b}'/[s, s+r_1]$$

which immediately implies

(2) $$\mathfrak{a}/[t, t+r_1] = \mathfrak{b}/[s, s+r_1],$$

where $\mathfrak{a}$ and $\mathfrak{b}$ are contractions of $\mathfrak{a}'$, $\mathfrak{b}'$, respectively, in $\mathscr{X}$. But by the assumptions concerning $\mathscr{X}$, (2) implies that

$$\mathfrak{a}/[t, t+r_0] = \mathfrak{b}/[s, s+r_0]$$

and hence we have

$$\mathfrak{a}'(t+r_0) = \mathfrak{b}'(s+r_0),$$

which contradicts (1) and concludes the proof.

2.5. DEFINABILITY

This section is to bridge considerations conducted thus far on closed systems with those which in the subsequent sections but the next one will be related to open systems. It establishes also certain links between dynamic regularities and regularities of structural type.

Let us begin with certain notational conventions. All systems to be considered in this section will be of the form:

(1) $$\mathfrak{a} = (a, F, G_1, \ldots, G_m, H_1, \ldots, H_n)$$

or, equivalently, of the form:

(2) $$\mathfrak{a} = (F^{\mathfrak{a}}, G_1^{\mathfrak{a}}, \ldots, G_m^{\mathfrak{a}}, H_1^{\mathfrak{a}}, \ldots, H_n^{\mathfrak{a}}).$$

The following symbols will be introduced to denote some of reducts of $\mathfrak{a}$:

(3) $$\mathfrak{a}^F = (F^{\mathfrak{a}}),$$

(4) $$\mathfrak{a}^G = (G_1^{\mathfrak{a}}, \ldots, G_m^{\mathfrak{a}}).$$

We shall study certain relations which variable F may bear to variables $G_1, \ldots, G_m$ or, equivalently, systems of the form $\mathfrak{a}^F$ may bear to systems of the form $\mathfrak{a}^G$.

DEFINITION 2.5.1. *Given a phenomenon* $\mathscr{Z}$ *whose instances are systems of the from* (1) *we shall say that* the values of F are definable in terms of the values of $G_1, \ldots, G_m$ (in $\mathscr{Z}$) *if and only if the following is satisfied*:

(Δ) *For any two systems* $\mathfrak{a}$, $\mathfrak{b}$ *in* $\mathscr{Z}$ *and for any instants* t, s,

$$F^{\mathfrak{a}}(t) = F^{\mathfrak{b}}(s)$$

whenever

$$G_i^{\mathfrak{a}}(t) = G_i^{\mathfrak{b}}(s), \quad i = 1, \ldots, m.$$

Any empirical law of the form

$$F = \Phi(G_1, \ldots, G_n), \tag{5}$$

where $\Phi(G_1, \ldots, G_n)$ is a formula built of parameters G_i and mathematical operations defined on the values of these parameters, provides an example of the regularity under discussion. The value of gravity with which a body a affects a body b is determined by the values of mass of the bodies and by the distance between them. The angle of refraction of a light ray which passes from one homogeneous medium into another is uniquely determined by velocity of light propagation in those two media.

In a general case of equality (5), mathematical operations used for constructing the formula $\Phi(G_1, \ldots, G_n)$ need not refer to the values of the parameters G_i but may refer directly to the parameters themselves. An example of such an operation is differentiation. As one may easily see, for instance, the values of velocity $\mathbf{v}$ are not defined in terms of the position s although by definition these two quantities are correlated as follows:

$$\mathbf{v} = \frac{ds}{dt}.$$

Still the values of acceleration a (being defined as the second derivative of s) are definable in terms of the values of s. Clearly the converse relations do not hold either: the values of s are neither definable in the values of $\mathbf{v}$ nor in the values of a. Observe however that whenever the values of one of these three parameters are known within any proper interval of time, the values of the others within that interval are easy to calculate (the values of s are calculable up to the initial position). This observation motivates the following definition.

DEFINITION 2.5.2. *Given a phenomenon $\mathscr{Z}$ whose instances are systems of the form* (1) *we shall say that F is* definable in terms of $G_1, \ldots, G_m$ (in $\mathscr{Z}$) within any proper interval of time *if and only if the following condition is satisfied*:

(Ξ) *For any two systems* $\mathfrak{a}$, $\mathfrak{b}$ *in* $\mathscr{Z}$ *and for any time intervals* u, v *of the same length,*

$$\mathfrak{a}^F/u =_\tau \mathfrak{b}^F/v,$$

whenever

$$\mathfrak{a}^G/u =_\tau \mathfrak{b}^G/v.$$

On comparing Conditions (Δ) and (Ξ) of the two definitions we immediately have

COROLLARY 2.5.3. If for a phenomenon $\mathscr{Z}$ the values of F are definable in terms of the values of $G_1, \ldots, G_m$, then F is definable in terms of $G_1, \ldots, G_m$ within any proper interval of time.

2.6. ONTOLOGICAL VERSUS SEMANTIC DEFINABILITY

The two concepts of definability we have discussed in the preceding section may be called ontological, because they refer to variables not to symbols by means of which the variables are spoken--about. It is of considerable interest to know how these notions are related to the notions of definability which possess a linguistic character.

DEFINITION 2.6.1. *Let $F, G_1, \ldots, G_m$ be among symbols of variables of the language of a theory Θ. We say that F is* definable (in semantic sense) in terms of $G_1, \ldots, G_m$ (in Θ) *if and only if the following condition holds true*:

(Σ) *For any two models* $\mathfrak{m}$, $\mathfrak{n}$ *for* Θ,

$$F^{\mathfrak{m}} = F^{\mathfrak{n}}$$

whenever

$$G_i^{\mathfrak{m}} = G_i^{\mathfrak{n}}, \qquad i = 1, \ldots, n.$$

(Clearly $F^{\mathfrak{m}}$, $F^{\mathfrak{n}}$, and so on stand for denotations which these symbols acquire under interpretations $\mathfrak{m}$ and $\mathfrak{n}$, respectively).

Note that condition (Σ) can be stated in an equivalent form as follows:

(Σ') *For any two models $\mathfrak{m}$, $\mathfrak{n}$ for Θ,*

$$\mathfrak{m}^F = \mathfrak{n}^F$$

whenever

$$\mathfrak{m}^G = \mathfrak{n}^G.$$

The way in which the notion of definability defined presently can be compared with those defined earlier is obvious. Denote by $Mod(\Theta)$ the set of realizations for Θ. The class $Mod(\Theta)$ plays with respect to $F, G_1, \ldots, G_m$ exactly the same role as it has been played by $\mathscr{X}$.

THEOREM 2.6.2. *Let $F, G_1, \ldots, G_m$ be among symbols of variables of a theory Θ. Denote by $Mod_{/\Sigma}(\Theta)$ the set of reducts of systems in $Mod(\Theta)$ to the variables $F, G_1, \ldots, G_m$. If:*

(C_1) *$Mod_{/\Sigma}(\Theta)$ is closed under translations in time, i.e. if $\mathfrak{m}$ is in $Mod_{/\Sigma}(\Theta)$ and $\mathfrak{m}'$ coincides up to a translation in time with $\mathfrak{m}$, then $\mathfrak{m}'$ is in $Mod_{/\Sigma}(\Theta)$.*

(C_2) *$Mod_{/\Sigma}(\Theta)$ is closed under restrictions in time, i.e. for each paper time interval u, $\mathfrak{m}/u$ is in $Mod_{/\Sigma}(\Theta)$ whenever $\mathfrak{m}$ is in $Mod_{/\Sigma}(\Theta)$,*

then the following two conditions are equivalent:

(i) *F is definable in terms of $G_1, \ldots, G_m$ in $Mod(\Theta)$ within any proper interval of time.*
interval of time.

(ii) *F is definable in terms of $G_1, \ldots, G_m$ in Θ.*

PROOF. The proof of this theorem is straightforward. Indeed if (i) holds true, then in view of the fact that Condition (Σ) of Definition 2.6.1. is entailed by Condition (Ξ) of Definition 2.5.2.

which is especially readily seen when the former is stated as (Σ'), (ii) must also be true. Thus in fact (i) entails (ii) quite independently of Assumptions (C_1), (C_2). The two clauses however are necessary in order to prove that (ii) implies (i). Let us assume that $\mathfrak{m}_1$, $\mathfrak{m}_2$ are in $Mod_{/\Sigma}(\Theta)$. It follows from (C_2) that for any two proper intervals u, v, the restrictions $\mathfrak{m}_1/u, \mathfrak{m}_2/v$ are also in $Mod_{/\Sigma}(\Theta)$. Assume that

(1) $$\mathfrak{m}_1^G/u =_\tau \mathfrak{m}_2^G/v.$$

Let $\mathfrak{m}$ coincide up to τ with $\mathfrak{m}_2/v$; note that by (C_1) $\mathfrak{m}$ is in $\mathrm{Mod}_{/\Sigma}(\Theta)$. We have then in particular

(2) $$\mathfrak{m}_2^G/v =_\tau \mathfrak{m}^G.$$

But the way in which $\mathfrak{m}$ has been defined yields that

(3) $$\mathfrak{m}_1^G/u = \mathfrak{m}^G.$$

Applying now (Σ') we arrive at

(4) $$\mathfrak{m}_1^F/u = \mathfrak{m}^F,$$

which, by the fact that $\mathfrak{m}_2/v =_\tau \mathfrak{m}$, entails

(5) $$\mathfrak{m}_2^F/v =_\tau \mathfrak{m}^F.$$

As an immediate consequence of (4) and (5) we obtain

(6) $$\mathfrak{m}_1^F/u =_\tau \mathfrak{m}_2^F/v,$$

which proves that (1) implies (6), establishing (Ξ) and concluding the proof.

One may suggest that semantic definability would turn out to be equivalent to definability in the sense of Definition 2.5.1. provided that clause (C_1) of Theorem 2.6.2. were not restricted to proper intervals, but required $Mod_{/\Sigma}(\Theta)$ to be closed under restrictions to point intervals (i.e. intervals of the form $[t, t]$) as well. From the practical point of view, however, interpretations which involve point intervals of time are of no interest unless they are treated as structures which do not involve time at all. If the notion of time is to play any role whatsoever, the durations of systems considered must be proper intervals. For that reason we shall assume that if a theory involves the notion of time, then it does not admit point-interval interpretations, i.e. such structures are not possible interpretations of the language of the theory. Incidentally, were such interpretations admitted, some of the mathematical operations (such as e.g. differentiation or integration with respect to time) would become meaningless.

Theorem 2.6.2. is evidently related to the problem of modelling of empirical regularities by means of suitable mathematical equations; we have discussed this problem in relation to state-determinism in Section 2.3.

Given a phenomenon $\mathscr{X}$ composed of systems of the form (1), denote by $C_{\mathscr{X}}$ a function (provided that it exists) of the form

(7) $$C_{\mathscr{X}}: St(\mathscr{X}^G) \rightarrow St(\mathscr{X}^F),$$

which satisfies the condition:

(*) For each $\mathfrak{a}$ in $\mathscr{X}$, and for each instant t

(8) $$C_{\mathscr{X}}(G_1(t), \ldots, G_m(t)) = G(t).$$

If such a function $C_{\mathscr{X}}$ exists, and clearly it exists if and only if the values of F are definable in terms of the values of $G_1, \ldots, G_m$, we shall call it a *state-correlation function* for $\mathscr{X}$, or shortly a *state-correlation*.

In a similar way one may define a history correlation $H_{\mathscr{X}}$ which would assign to each history $\mathfrak{a}^G/u$ the corresponding history $\mathfrak{a}^F/u$, provided that such a correspondence is defined uniquely in $\mathscr{X}$, i.e. F is definable in terms of $G_1, \ldots, G_m$ within any proper interval. Let us restrict our discussion to the function $C_{\mathscr{X}}$, however.

The conditions by means of which the notion of a mathematical model should be defined are obvious analogons of the conditions (+) and (++). We shall say then that a set of equations E stated in a suitable language is a *mathematical model* of $C_{\mathscr{X}}$ provided that the following two conditions are satisfied:

(*) the values of F are defined in terms of the values of $G_1, \ldots, G_m$ in $Mod(E)$,

(**) $\mathscr{X} \subseteq Mod(E)$.

Clearly we have the following

THEOREM 2.6.3. *If* $\mathscr{X}$ *and* Θ *are as assumed above and there exists a mathematical model E of* $C_{\mathscr{X}}$ *in* Θ, *then F is definable (in the semantic sense) in terms of* $G_1, \ldots, G_m$ *in* Θ.

PROOF. By the assumptions of the theorem the values of F are defined in terms of the values of $G_1, \ldots, G_m$ in $Mod(E)$ and $E \subseteq \Theta$. The latter fact implies that $Mod(\Theta) \subseteq Mod(E)$, and thus, as one may easily verify, the values of F are defined in terms of the values of $G_1, \ldots, G_m$ in Θ as well. Suppose now that $\mathfrak{m}$, $\mathfrak{n}$ are in $Mod(\Theta)$ and that $\mathfrak{m}^G = \mathfrak{n}^G$. Then clearly for each t, we have $G_i^{\mathfrak{m}}(t) = G_i^{\mathfrak{n}}(t)$, $i = 1, \ldots, m$. This however implies $F^{\mathfrak{m}}(t) = F^{\mathfrak{n}}(t)$, for all t, and establishes the theorem.

2.7. SURROUNDING CONDITIONS

Thus far we have tacitly assumed that the systems considered are not subject to any actions from outside which might disturb their way of behaviour. That not necessarily should be interpreted as equivalent to the assumption that the systems were taken to be isolated from their environments. From the formal standpoint it is irrelevant whether a system is protected against any influences from outside or whether influences which might affect the system do not exist although the "channels of communication" between the system and its neighbourhood are in order, i.e. an exchange of energy or/and information is possible. These casual remarks can be easily stated in a more precise way.

Let $\mathscr{X}$ be a phenomenon consisting of open systems of the form

(1) $$\mathfrak{a} = (F_1, \ldots, F_m, G_1, \ldots, G_n),$$

the variables F_i being inputs and the variables G_i being outputs of $\mathfrak{a}$. Let us put

(2) $$\mathfrak{a}^F = (F_1, \ldots, F_m),$$

(3) $$\mathfrak{a}^G = (G_1, \ldots, G_n).$$

The constraint $\mathfrak{a}^F$ will be referred to as an *input system* and $\mathfrak{a}^G$ as an *output system*, $\mathfrak{a}^F$ being interpreted as the *environment* of $\mathfrak{a}^G$.

Denote by $\mathscr{X}^F$ the set of all input systems $\mathfrak{a}^F$, where $\mathfrak{a}$ belongs to $\mathscr{X}$, and similarly, denote by $\mathscr{X}^G$ the set of all output systems corresponding to the systems in $\mathscr{X}$. Each subset $\mathscr{W} \subseteq St^*(\mathscr{X}^F)$ will be referred to as (*possible*) *surrounding conditions* (for $\mathscr{X}^G$).

DEFINITION 2.7.1. *A phenomenon $\mathscr{X}$ whose instances are open systems of the form* (1) *is* **state-determined under surrounding conditions** $\mathscr{W}$ *if and only if for all time intervals u, for all* $\mathfrak{a}$, $\mathfrak{b}$ *in $\mathscr{X}$, for all instants t_1, t_2 in u, and for all r such that both t_1+r and t_2+r are in u, if*

(i) $\mathfrak{a}^G(t_1) = \mathfrak{b}^G(t_2)$,

then

(ii) $\mathfrak{a}^G(t_1+r) = \mathfrak{b}^G(t_2+r)$,

provided that for all s in u, $\mathfrak{a}^F(s) \in \mathscr{W}$.

Definition 2.7.1. is an obvious generalization to Definition 2.2.2. As it is readily seen the following is an immediate consequence of the two definitions.

COROLLARY 2.7.2. *Let $\mathscr{X}$, $\mathscr{X}^F$, $\mathscr{X}^G$ be as defined above. Then $\mathscr{X}^G$ is state-determined if and only if $\mathscr{X}$ is state-determined under any surrounding conditions (or equivalently, under surrounding conditions $St^*(\mathscr{X}^F)$).*

Clearly we also have

COROLLARY 2.7.3. *Let $\mathscr{X}$, $\mathscr{X}^F$, $\mathscr{X}^G$ be as thus far. Given any surrounding conditions $\mathscr{W}$, define $\mathscr{X}^G_{\mathscr{W}}$ as the set of all systems of the form $\mathfrak{a}^G/u$ which satisfy the following conditions:*

(i) $\mathfrak{a}$ *is in* $\mathscr{X}$,

(ii) *for each t in u, $\mathfrak{a}^F(t)$ is in $\mathscr{W}$,*

(iii) *for each interval v such that $u \subset v \subseteq T(\mathfrak{a})$ there exists an instant s such that $\mathfrak{a}^F(s) \notin \mathscr{W}$.*

Under these assumptions $\mathscr{X}^G_{\mathscr{W}}$ is state-determined if and only if $\mathscr{X}$ is state-determined under conditions $\mathscr{W}$.

If $St^*(\mathscr{X}^F)$ can be partitioned into disjoint subsets $\mathscr{W}_1, \ldots, \mathscr{W}_k$ such that $\mathscr{X}$ is state-determined under each of surrounding conditions $\mathscr{W}_i$ and at the same time the state-transformation D_t for corresponding phenomena $\mathscr{X}_{\mathscr{W}_i}$ differ from each other, we are inclined to treat $\mathscr{X}$ as a conglomerate of different phenomena $\mathscr{X}_{\mathscr{W}_i}$ rather than as a single phenomenon in the intuitive sense of the word (cf. the discussion in Section 1.6 concerning possible ways of defining the phenomenon TB_V).

Let us now discuss the case, much more involved, when different states **x** in $St(\mathscr{X}^F)$ affect in different ways the states of the systems in $\mathscr{X}^G$, and thus each change in the environment of a system $\mathfrak{a}^G$ either causes a change of the state of that system or leads to a change of the way in which the states of the system transform in time or has both effects.

The strongest relation which outputs may bear to inputs is that of definability. In the extreme case of system-surrounding dependence, the values of output variables may turn out to be defined by the values of input variables. If such a regularity takes place, the state-correlations between input and output variables do not define by themselves the way in which output variables change their values in time. But those changes become fully determined when the changes of input variables are defined. By no means this form of dependence should be treated as a typical one. In fact it is rather exceptional. Practically in most cases when the behaviour of a system is determined by changes in the environment, it is determined up to certain initial conditions. In the simplest case those conditions are defined provided that the state of the system at an instant (usually an arbitrary one) is given. That state is then referred to as the *initial state* of the system.

DEFINITION 2.7.4. *We shall say that a phenomenon $\mathscr{X}$ whose elements are open systems of the form* (1) *is* input-determined *if and only if for all* $\mathfrak{a}$, $\mathfrak{b}$ *in* $\mathscr{X}$ *and for all t, s, the following condition is satisfied*:

(*) *If* $\mathfrak{a}^F(t) = \mathfrak{b}^F(s)$, *then* $\mathfrak{a}^G(t) = \mathfrak{b}^G(s)$,

provided that for some t_0, s_0

$$\mathfrak{a}(t_0) = \mathfrak{b}(s_0).$$

In a loose language we can recast this definition as follows: $\mathscr{X}$ is *input-determined* if and only if the state of each system

$\mathfrak{a}^G$ in $\mathscr{Y}^G$ at each moment t is determined uniquely by the state of its surrounding $\mathfrak{a}^F$ and the initial states of the coupled systems $\mathfrak{a}^G$, $\mathfrak{a}^F$.

An easy argument allows to establish the following:

ASSERTION 2.7.5. *If $\mathscr{Y}$ is input-determined and the systems in $\mathscr{Y}$ have a common (initial) state, i.e. there exists a state $\mathbf{z} \in St(\mathscr{Y})$ such that $\mathbf{z} \in St(\mathfrak{a})$, for all $\mathfrak{a}$ in $\mathscr{Y}$, then the value of each output variable of $\mathscr{Y}$ is definable in terms of the values of input variables.*

The requirement that $\mathfrak{a}(t_0) = \mathfrak{b}(s_0)$, i.e. the requirement that $\mathfrak{a}$ and $\mathfrak{b}$ have certain common initial state may happen to be too weak to guarantee that condition (*) of Definition 2.7.4. is satisfied. Clearly in that case $\mathscr{Y}$ is not input-determined in the sense defined, but it may still be „input-determined" in the following weaker sense: Condition (*) holds true provided that certain parts of histories of $\mathfrak{a}$ and $\mathfrak{b}$ coincide.

To have an example when initial conditions are defined by a certain fragment of a history of a system rather than by a singular state consider any system whose surrounding conditions have undergone a drastic and abrupt change. Practically all empirical systems, regardless of what kind they are (political, economical, ecological, mechanical, chemical, psychological, etc, etc), require a period of adaptation before they begin to behave in a "normal" way under new surrounding conditions, i.e. in the way which falls under regularities typical for those conditions. Clearly, in order to decide whether the change is abrupt or not one has to study a part of history of a pertinent input system. In order to distinguish between the two kinds of input-determined phenomena, i.e. those being input-determined in the sense of Definition 2.7.4. and those being such in the sense of our informal remarks, one may call the former *state-input-determined* and the latter *history-input-determined.*

In all those cases when the behaviour of a system is determined by certain initial conditions and changes which take place in its environment, we shall say that the environment dominates the system. We shall not treat the term "domination" as a technical one, however.

2.8. SELF-DETERMINED PHENOMENA

Of considerable interest are those phenomena whose instances are systems with feedback, i.e. the systems which do not undergo changes extorted by their environments passively but react back. In the ultimate case the systems may entirely control the state of their environments.

It is rather obvious that in order to describe the domination of a system over its environment we merely have to reverse the rules played by input and output systems in our earlier discussion. Thus in particular we shall say that a phenomenon $\mathscr{X}$ is *output-determined* if the definiens of Definition 2.7.4. holds true, with the proviso that $G_1, \ldots, G_n$ are now defined to be input and $F_1, \ldots, F_m$ output variables. A phenomenon which is both input and output determined will be said to be self-determined. Clearly this definition amounts to the following

DEFINITION 2.8.1. *We shall say that a phenomenon $\mathscr{X}$ whose elements are open systems of the form defined in the preceding section by the formula* (1) *is* self-determined *if and only if for all* $\mathfrak{a}$, $\mathfrak{b}$ *in* $\mathscr{X}$ *and for all t, s,*

$$\mathfrak{a}^F(t) = \mathfrak{b}^F(s) \text{ if and only if } \mathfrak{a}^G(t) = \mathfrak{b}^G(s),$$

provided that for some t_0, s_0

$$\mathfrak{a}(t_0) = \mathfrak{b}(s_0).$$

It is easily seen that a phenomenon which is self-determined need not be state-determined. Assume for instance that the conceptualization of $\mathscr{X}$ consists of exactly two variables F and G, F being an input and G an output variable. Let for each $\mathfrak{a}$ in $\mathscr{X}$ and for each t

$$F(t) = G(t+\pi).$$

At the same time assume that for each $\mathfrak{a}$ in $\mathscr{X}$ there exists a $t_{\mathfrak{a}}$ such that

$$G(t) = \sin(t+t_{\mathfrak{a}}).$$

One may easily verify that $\mathscr{X}$ is self-determined, but is not state-determined. (Incidentally, the example given shows also that periodic systems in the sense defined in Assertion 2.4.5 need not be state-determined).

The converse need not be valid either. If $\mathscr{X}$ is state-determined there need not exist a partition of the conceptualization of the phenomenon onto input and output variables such that $\mathscr{X}$ is self-determined with respect to this partition.

In spite of the fact that the notions we have compared are distinct, there is an apparent affinity between them. Let us examine the matter closer.

In order for a phenomenon $\mathscr{X}$ to be uniquely self-determined for each $\mathfrak{a}$ in $\mathscr{X}$ there must exist two functions f and g such that for each t,

(1) $$\mathbf{y} = f(\mathbf{x})$$

and

(2) $$\mathbf{x} = g(\mathbf{y}),$$

where $\mathbf{x} = \mathfrak{a}^F(t)$ and $\mathbf{y} = \mathfrak{a}^G(t)$. The existance of such functions is an obvious consequence of Clause (iii) of Definition 2.8.1.

Taking into account that

$$\mathbf{x} = (x_1, \ldots, x_m)$$

and

$$\mathbf{y} = (y_1, \ldots, y_n)$$

we can replace (1) and (2) by the following two systems of equations:

(3) $$y_i = f_i(x_1, \ldots, x_m), \quad i = 1, \ldots, n,$$

and

(4) $$x_i = g_i(y_1, \ldots, y_n), \quad i = 1, \ldots, m,$$

or equivalently by

(5) $$G_i(t) = f_i(F_1(t), \ldots, F_m(t)), \quad i = 1, \ldots, n,$$

and

(6) $$F_i(t) = g_i\big(G_1(t), \ldots, G_n(t)\big), \quad i = 1, \ldots, m,$$

respectively.

The latter presentation brings to light the dependence of arguments of the equations on time. In practical situations, differential counterparts of (5) and (6) (or (3), (4)) are most likely to be easier available because when trying to describe the behaviour of the system, one almost always compares changes of different parameters relevant to the problem.

Differential counterparts of equations (5), (6), if they exist, usually take the following form:

$$\frac{dG_i}{dt} = a_{i1}\frac{dF_1}{dt} + \ldots + a_{in}\frac{dF_m}{dt}, \quad i = 1, \ldots, m,$$

$$\frac{dF_i}{dt} = b_{i1}\frac{dG_1}{dt} + \ldots + b_{in}\frac{dG_n}{dt}, \quad i = 1, \ldots, n,$$

where the coefficients a_{ij} and b_{ij} are defined as follows:

$$a_{ij} = \frac{\partial g_i}{\partial x_j},$$

and

$$b_{ij} = \frac{\partial f_i}{\partial y_j}.$$

It is obvious that in order for a self-determined phenomena to be state-determined, equations (5) and (6) or any of their counterparts must have a unique solution, provided that initial conditions were stated. In general such a unique solution need not exist. None the less even if equations (5), (6) do not suffice to predict in a unique manner the changes which the systems of the phenomenon $\mathscr{X}$ may undergo, still they may provide us with a considerable amount of information about the behaviour of those systems.

To appreciate how powerful the analysis of Sem-environment interactions can be even if it does not result in the discovery of any strictly deterministic regularity, let us consider a few examples. To make the thing simpler let us pass over to discrete time approach, i.e. let us assume that the phenomena we deal with are subject to observations only at certain distinguished instants which form a sequence $\{t_n\}$.

We shall say that $\mathscr{X}$ is *self-determined with respect to* $\{t_n\}$ if and only if the definiens of Definition 2.8.1. holds true when the time instants to which it refers are selected from $\{t_n\}$. This selfexplanatory decision immediately yields the following:

ASSERTION 2.8.2. *$\mathscr{X}$ is self-determined with respect to $\{t_n\}$ if and only if there exist two functions λ and ϱ such that the following conditions are satisfied*:

(i) $\lambda\colon St(\mathscr{X}) \times St(\mathscr{X}^G) \to St(\mathscr{X}^F)$,

(ii) $\varrho\colon St(\mathscr{X}) \times St(\mathscr{X}^F) \to St(\mathscr{X}^G)$,

and, for each $\mathfrak{a}$ *in* $\mathscr{X}$ *and for each* t_i:

(iii) $\lambda(\mathfrak{a}(t_i), \mathfrak{a}^G(t_{i\pm1})) = \mathfrak{a}^F(t_{i\pm1})$,

(iv) $\varrho(\mathfrak{a}(t_i), \mathfrak{a}^F(t_{i\pm1})) = \mathfrak{a}^G(t_{i\pm1})$.

(It is worthwhile noticing that while the notion of state-determination with respect to a discrete sequence $\{t_n\}$ corresponds to the notion of deterministic automaton as defined by M. O. Rabin and D. Scott (1959), the notion of self-determination corresponds (when related to a finite sequence) to the notion of an automaton as defined for instance in the monograph by R. E. Kalman, P. L. Falb and M. A. Arbib (1969)).

Each instance $\mathfrak{a}$ of a phenomenon $\mathscr{X}$ being self-determined with respect to $\{t_n\}$ can be viewed as a two-player game. The initial state of the system corresponds to the initial position in the game, and each of the successive positions (states) results from successive moves executed by the players. Let us denote the two players as F and G; the former will be assumed to control the inputs, the latter the outputs of $\mathfrak{a}$. At no stage of the game we can predict who will make the next move and what move will be made unless $\mathscr{X}$ is self-determined. But we know exactly how any of the two players will react to any change of the position produced by the other.

The important thing is that if the aims which the players want to achieve are known we may happen to be able to say a lot about possible outcomes of the game and even predict the final result. What we have to do for that purpose is to analyse whether any of the two players has a *winning strategy*, i.e. whether there exists a function $\sigma_F\colon St(\mathscr{Y}^F) \to St(\mathscr{Y}^F)$ or a function $\sigma_G\colon St(\mathscr{Y}^G) \to St(\mathscr{Y}^G)$ such that the player F must win (i.e. achieve its aim) if the moves he executes are successively:

$$\sigma_F(\mathfrak{a}^F(t_0)),\ \sigma_F(\sigma_F(\mathfrak{a}^F(t_0))), \ldots$$

or correspondingly G must win if he applies in the same sense the strategy σ_G.

How much we are able to predict after discovering that a winning strategy exists depends on circumstances. If a fox chases a rabbit the goals are obvious: the fox wants to catch the prey, the rabbit wants to escape the predator. Even if we know that, say, the rabbit has a winning strategy, it does not mean that the poor creature will be able to discover it.

Curiously enough the most interesting case seems to be not that of existence of a winning strategy but that when each of the two players has the best strategy in the sense that if either player applies his best strategy, he cannot lose the game. Consequently, if both the players apply their best strategies the game must result in a "draw".

Assume, for instance, that a population of rabbits and that of foxes live on the same territory, and rabbits are the only source of food for foxes. The "goals" of the two populations are again clear: maximal increase in number. Neither of the two "opponents" has a winning strategy however, the foxes cannot eat all rabbits if they are to survive. The amount of the food available limits also the number of rabbits.

If we state the conditions under which the problem is to be considered in a definite and precise manner, we shall be able to prove that depending on such parameters as: the appetite of rabbits, the amount of grass, bark, and what else they can eat, the daily ratio of rabbits per fox necessary for the latter to be in good health, and hunting abilities of foxes, there exists an optimal number n_r of rabbits and an optimal number n_f of foxes which can happily live together on the territory. These two numbers define the *state of equilibrium*, i.e. only if there is exactly n_r rabbits and n_f foxes at, say, the end of breeding season, their number at that time next year will be the same. Any surplus of rabbits over n_r will result in an increase of the number of foxes but this cannot last too long (there will not be enough food for rabbits). Any surplus of foxes over n_f cannot be stable either, again the regulating factor is the amount of food available (this time the number of rabbits). Thus if for one reason or another a deviation from the state of equilibrium has occurred, the whole system will tend through diminishing oscillations to the state of equilibrium (For a rigorous discussion of the problem cf. J. M. Smith, *Mathematical Ideas in Biology*, 1968).

The general idea of a mathematical technique of looking for the state of equilibrium (provided that it exists) is simple. Given equations (5), (6) (or any other equivalent to them) we have to try to solve those equations together with the additional ones of the form

$$F_i(t) = c_i, \quad i = 1, \ldots, m,$$
$$G_i(t) = d_i, \quad i = 1, \ldots, n,$$

where c_i and d_i are constants (i.e. they do not depend on time).

The game terminology adopted here mainly for didactic purposes need not be treated as out of place when the whole discussion is to be conducted in a rigorous fashion. Quite to the contrary, starting from the classical work *Theory of Games and Economic Behaviour* by J. von Neumann and O. Morgernstern theory of games serves as a powerful instrument of analysing conflict tendencies in all areas of human activity. The techniques of theory of games can be applied also to analysis of animal's behaviour. Even such processes as evolution of species can be viewed as processes of realizations of "evolutionary stable strategy" (cf. J. M. Smith and G. R. Price, 1973) and what is more, even inanimate processes can occasionally be metaphorically treated as processes of selfdefence of systems against changes in their surroundings. For instance, thermodynamic systems react "as if they tended to minimize the effect of such changes" (*Le Chatelier's principle*).

Clearly the mathematical method by means of which interactions among coupled systems (pairs of such systems or systems being parts of a complex one) can be studied need not necessarily involve the concepts specific for theory of games.

Behaviour of complex systems, in particular criteria for their stability, has been subject to intensive investigations conducted with the help of various and in most cases extremely sophisticated techniques.

2.9. INVARIANCY

Thus far we have not treated time as a pair with other variables of systems. This has been motivated both by convenience of such an approach and the fact that time (note that the same might be said about space) plays a special role in empirical investigations. Outside of physics, more accurately a certain part of it, time is only an auxiliary parameter, although of a great significance; it is not an object of inquiry by itself.

For the purpose of considerations which are to be conducted in this section it becomes desirable to treat time as an additional variable of the systems under discussion. The easiest, although indirect, way in which that can be done is to pass on to the phase-space approach.

To begin with let us introduce certain terminological conventions. Let $\mathscr{E}^{n+1}$ be the phase-space of $\mathscr{X}$. The elements (points) of $\mathscr{E}^{n+1}$ are then couples of the form $(\mathbf{x}, t)$, where $\mathbf{x} \in St(\mathscr{X})$ and $t \in Re$. We shall refer to them as *events* (cf. Section 1.7). Each event can be interpreted as a (possible) state of systems in $\mathscr{X}$ expanded by adding time as a new explicit variable. This is then exactly what we wanted to have.

If an event $\mathfrak{x}$ is of the form $(\mathfrak{a}(t), t)$, where $\mathfrak{a}$ is in $\mathscr{X}$, we shall say that $\mathfrak{x}$ *occurs on* $\mathfrak{a}$. Since from the formal standpoint $\mathfrak{a}$ can be identified with its line of behaviour, consequently the events which occur on $\mathfrak{a}$ can be treated simply as elements of $\mathfrak{a}$.

Let Γ be a group of one-to-one transformations of $\mathscr{E}^{n+1}$ onto itself. The word "group" should be understood here in its mathematical sense, the composition of two transformations being taken as the group operation and the identity transformation as the unit of the group. Thus the following conditions should be satisfied:

(Γ_1) The identity transformation (i.e. the transformation i such that for each $\mathfrak{x} \in \mathscr{E}^{n+1}$, $i(\mathfrak{x}) = (\mathfrak{x})$ is in Γ.

(Γ_2) If f is in Γ, then f^{-1} is in Γ.

(Γ_3) If f and g are in Γ, then the composition $f \circ g$ is in Γ. (the conditions are not independent since ($\Gamma 1$) is an immediate consequence of the remaining two. Indeed $i = f \circ f^{-1}$, for any f in Γ).

Let $f \in \Gamma$. Given any system $\mathfrak{a}$ in $\mathscr{X}$ we shall write $f(\mathfrak{a})$ to denote the image of $\mathfrak{a}$ by f, i.e.

$$f(\mathfrak{a}) = \{f(\mathfrak{x}) : \mathfrak{x} \in a\}.$$

We also put:

$$f(\mathscr{X}) = \{f(\mathfrak{a}) : \mathfrak{a} \in \mathscr{X}\}.$$

Since we identify systems with their lines of behaviour, if $\mathfrak{a}$ is a system, then $f(\mathfrak{a})$ is a system again and thus $f(\mathscr{X})$ is a set of systems similar to those in $\mathscr{X}$. These, as a matter of fact, obvious remarks should make the following definition intelligible

DEFINITION 2.9.1. *Let $\mathscr{X}$ be a phenomenon,* **R** *a regularity of $\mathscr{X}$ (i.e. $\mathscr{X} \in \mathbf{R}$) and let Γ be a group of transformations of the phase space $\mathscr{E}^{n+1}$ of $\mathscr{X}$. We shall say that* **R** *is preserved on $\mathscr{X}$ under Γ if and only if for all $f \in \Gamma$, $\mathscr{X} \cup f(\mathscr{X}) \in \mathbf{R}$.*

To give an example of application of the concept defined assume that the states $\mathfrak{a}(t)$, $\mathfrak{b}(s)$ of the systems $\mathfrak{a}$, $\mathfrak{b}$ of particles differ only with respect to the positions occupied by the particles being the elements of the two systems, but still these positions coincide up to certain translations in the physical space. If the position is one of the variables by means of which we describe the systems, then we obviously have:

$$\mathfrak{a}(t) \neq \mathfrak{b}(s).$$

At the same time however, provided that f is defined to be a suitable translation in space, we have

$$\mathfrak{a}(t) = f(\mathfrak{b}(s)),$$

and the laws of mechanics guarantee that for all r

$$\mathfrak{a}(t+r) = f(\mathfrak{b}(s+r)).$$

Thus the systems which observe laws of particle mechanics are not only determined but their state-determined behaviour is preserved under translations in the physical space. We have presented this example in a loose manner but one can easily recast it in precise terms.

There are much more transformations which are considered by physicists as not affecting physical laws. For example not only translations but also turns, and mirror deflections belong to this category. For a wider discussion of invariancy of physical laws (and thus also regularities they describe) under certain translations cf. e.g. R. P. Feynman, R. B. Leighton and R. B. Sands (1966).

The invariancy of a regularity under certain transformations is a property of that regularity and consequently it is a "second order" property of those phenomena which display the regularity and on which regularity is preserved. In fact then such an invariancy is a regularity by itself.

Observe that all the regularities we have defined in the preceding section are by definitions invariant with respect to translations in time. Indeed, let $\mathscr{E}^{n+1}$ be the phase-space of $\mathscr{Z}$. A translation in time is then a mapping

$$\tau_r : \mathscr{E}^{n+1} \to \mathscr{E}^{n+1}$$

such that

$$\tau_r(x_1, \ldots, x_n, t) = (x_1, \ldots, x_n, t+r).$$

It is an easy matter to verify that, for instance, if $\mathscr{Z}$ is state-determined, then $\mathscr{Z} \in \tau_r(Z)$ is also state-determined.

Yet, it would be a mistake to conclude from what has been said above that the invariancy with respect to translations in time is a definitional property of regularities. Clearly, if a regularity is defined to involve invariancy with respect to time it does this but, for instance, the condition by means of which the notion of a state-determined phenomenon was defined (cf. Definition 2.2.2) might be replaced by the following:

(SD) For all $\mathfrak{a}$, $\mathfrak{b}$ in $\mathscr{X}$ and for all t, r ($r \geqslant 0$), if $\mathfrak{a}(t) = \mathfrak{b}(t)$, then $\mathfrak{a}(t+r) = \mathfrak{b}(t+r)$.

Each state-determined (in the sense of Definition 2.2.2) phenomenon is state-determined in the sense of Condition (SD) but not the other way round. We have merely the following:

ASSERTION 2.9.2. *A phenomenon $\mathscr{X}$ is state-determined if and only if it is state determined in the sense of Condition* (SD) *and the regularity which* (SD) *defines is preserved on $\mathscr{X}$ under translations in time.*

Clearly, in general, given a phenomenon $\mathscr{X}$ which has (SD)-*-property*, as we can say for short, we have no guarante that this property is on $\mathscr{X}$ invariant with respect to translation in time. If it is, this is an empirical fact which cannot be proved by purely formal considerations (cf. the discussion concerning the empirical character of invariancy with respect to translations in time in R. P. Feynman, R. B. Leighton and M. Sands (1966).

Let us conclude this section with a short discussion of an issue different in its character but nevertheless quite strongly related to the earlier considerations of this section. Thus far we have tacitly assumed that when different systems are compared by comparing their states at certain instants (we can now equivalently say: by comparing events which occur on them), the variables the systems involve are measured with respect to a fixed system of units and a fixed frame of reference.

This assumption may unfortunately happen to be unrealistic, especially as far as the notion of a frame of reference is concerned. Two micro-systems (in the physical sense of the word) which are far away from each other in space or in time cannot be in any reasonable way examined within the same frame of reference, if the latter is to be something more than a purely mathematical notion.

One may argue that although an operational definition of a common frame of reference for all systems regardless of their size, space they occupy, length of their durations etc. is impossible indeed, the notion of such a frame is not meaningless from a theoretical point of view. This is doubtful again, but even if we assume that the idea of such a frame is acceptable on purely theoretical grounds, still for practical reasons we must look for such an account of an empirical phenomenon which will enable us to compare the systems (at least some of them) which are defined with respect to different frames of reference.

Observe that both the system of units and the frame of reference can be treated as parameters which serve to distinguish different phase-spaces. Let us then introduce the symbol $\mathscr{E}^{n+1}_{\mu,\varrho}$ to denote the phase-space of those systems belonging to a given phenomenon $\mathscr{Z}$ which are defined with respect to the system of units μ and the frame of reference ϱ. Clearly if all systems in $\mathscr{Z}$ are defined with respect to the same μ and same ϱ, the indices μ, ϱ can be dropped out and we pass to the case we have dealt with in all earlier sections of this chapter.

In order to compare systems which belong to different phase--spaces, in particular in order to speak about regularities common to them, we must have to our disposal a group of transformations Γ of the form

$$f\colon \mathscr{E}^{n+1}_{\mu,\varrho} \to \mathscr{E}^{n+1}_{\nu,\sigma},$$

being *onto* and *one-to-one*. Clearly there is a great number of groups of transformations of this sort. But when such a problem arises in empirical investigation the group we are looking for has to have a certain specific property: the regularities which phenomena examined display should be left invariant under all transformations the group involves. Such a role, to give an obvious and immediate example, is played by the *Lorentz transformations* with respect to regularities of mechanical and electrodynamical phenomena.

The sense in which one may say that a regularity **R** of $\mathscr{Z}$ is preserved under the group Γ as defined above needs to be defined. To begin with let us denote by $\mathscr{Z}_{\mu\varrho}$ the subphenomenon of $\mathscr{Z}$ that comprises all systems of $\mathscr{Z}$ which are represented in the phase space $\mathscr{E}^{n+1}_{\mu\varrho}$. We shall attribute **R** to $\mathscr{Z}$ if and only if **R** is a regularity of all subphenomena of the form $\mathscr{Z}_{\mu\varrho}$. Thus for instance we shall say that $\mathscr{Z}$ is state-determined if and only if each $\mathscr{Z}_{\mu\varrho}$ is state-determined. The convention we have just adopted connects our earlier considerations with the present ones in a natural and simple way. It allows also to apply Definition 2.9.1 without any changes. We shall say that **R** is preserved on Z under Γ provided that $\mathscr{Z} \cup f(\mathscr{Z}) \in \mathbf{R}$, for all f in Γ. This explains the sense in which one may speak about invariancy of laws under transformations which involve changes of units or frame reference and concludes the considerations of this section.

2.10. NOTES

A. *The notion of determinism – historical remarks.* One of the clearest formulations of the notion of determinism is due to Laplace who in his "*Essai philosophique sur les probabilités*" wrote these famous words:

> "An intelligence which knows the forces acting in nature at a given instant, and the mutual positions of the natural bodies upon which they act, could, if it were furthermore sufficiently powerful to subject these data to mathematical analysis, condense into a single equation the motion of the largest heavenly bodies and of the lightest atoms; nothing would be uncertain for it, and the future as well as the past would lie open before its eyes. The human mind, in the perfection to which it has carried astronomy, offers a weak image of such an intelligence in a limited field."

While Laplace and many other thinkers attributed determinism to the Universe as a whole, nowadays we are acutely aware that the problem of whether the Universe is deterministic or not must be meaningless, since no criteria of adequate solution to this problem can be given. This is a typical metaphysical problem which can be subject of philosophical speculations and not of scientific investigations.

There are two main presentations in which the problem of

determinism appears in the current philosophical and methodological discussions. Those who are working in the field of theory of systems attribute the notion of determinism to systems and define the notion of a deterministic system in substantially the same way in which the notion of a state-determined phenomenon (Definition 2.2.2) has been defined. Quite independently the principle of determinism is sometimes postulated as a rule which should be observed in scientific activity. Let me quote in this connection the following remark by W. A. Ashby (1960, p. 28):

> "Because of its importance, science searches persistently for the state-determined. As a working guide, the scientist has for some centuries followed the hypothesis that, given a set of variables, he can always find a larger set that (1) includes the given variables, and (2) is state-determined. Much research work consists in trying to identify such a larger set, for when the set is too small, important variables will be left out of account, and the behaviour of the set will be capricious. The assumption that such a larger set exists is implicit in almost all science, but, being fundamental, it is seldom mentioned explicitly."

A different approach is represented by those theorists who attribute determinism to theories rather than to the Universe or any part of it. Thus the last approach might be called *linguistic* in contradistinction to the former which is *ontological* in its character. A deterministic theory is defined to be a theory which (let us quote E. Nagel, 1953, p. 422):

> "enables us given a state of (a system)... at one time to deduce the formulation of the state at any other time".

The controversy between ontological and linguistic approaches to the problem of determinism has a long history and I am not be able to reconstruct it in a satisfactory manner. Anyway, taking the linguistic standpoint, one arrives again at the notion of determinism which can possess a precise and rigorously defined content. With the proviso that the language of a theory is well--defined, we can define the theory to be deterministic if the class of realizations of the theory is state-determined. In this way we arrive at a semantic notion of the determinism of theories; clearly this notion can also be studied in a syntactical way. For a wide and illuminating discussion of the notion of determinism in application to Particle Mechanics cf. R. Montague, (1962). The problem of determinism of theories (in particular in application to econometry) has recently been studied by V. Rantala (1975). An extended discussion of various kinds of laws (and thus a discussion of different kinds of regularities from the linguistic point of view) can be found in a paper by F. Suppes (1976).

The approach adopted in this book is definitely ontological. If I had to justify my decision I would try to do this on a philosophical ground: fundamental properties of theories which we construct are determined by properties of phenomena, not the other way round. Thus in particular, I am inclined to believe that the determinism of theories (those which are deterministic) reflects certain properties of empirical systems which the theories describe. One of the tasks of philosophy of science is to account for the very nature of this dependence. I shall turn back to this problem, which is much more complicated than the relative simplicity of the correspondence between deterministic theories and deterministic systems may suggest, in the next chapter.

Let me conclude these remarks by commenting shortly on the decision to attribute regularities and in particular determinism to sets of systems rather than to single systems. Since the motives for this decision have been stated in a less or more explicit manner in Sections 1.6 and 2.1, the present remarks are mainly to complement what has been said earlier, and first of all to relate the standpoint taken in this book to the positions held in this matter by some other authors.

The most important thing, when the divergences between the single-system and set of systems (phenomenon) approach is discussed, is to realize that the differences may be of purely verbal nature. If for instance, W. A. Ashby discusses state-determination of systems, the systems with which he deals are what can be called a mathematical representation of empirical systems. The behaviours of a great many different physical objects, each being examined with respect to certain fixed, common for all of them, variables, can fall under the same pattern. In this way instead of studying those systems we can study the pattern of their behaviour, in particular we can pose the question of under what conditions this pattern may be said to be deterministic. Ashby's concept of a system is the formal counterpart of our informal concept of "pattern of behaviour".

It should be pointed out however that the idea to attribute the notion of determinism to classes of systems has occasionally been expressed in methodological discussion. Thus, for instance, R. Montague defined the concept of a deterministic class of histories (1962, Definition 11) although, as a matter of fact, he advocates very strongly the linguistic conception of determinism. In particular, when commenting on the notion of determinism suggested by Theorem 2 stated in McKinsey, Sugar and Suppes (1953), he writes: "General notions of determinism as applied to classes of models or histories seem to be of rather restricted interest" (cf. Montague, 1962, Footnote 18). As one may guess from this remark, McKinsey, Sugar and Suppes's concept of determinism applies also to classes of systems. Incidentally, Montague's disbelief in significance of the concept of determinism of class of systems seems to be an indication of his disaproval of an ontological conception of determinism in any version whatsoever.

B. *Causality*. The discussion of the concept of causality usually involves conceptual apparatus different from the one we have applied in the discussion carried out in this chapter. For a wide and rigorous discussion of that concept cf. the monograph by P. Suppes *A Probabilistic Theory of Causality* (cf. also A. W. Burks, 1951, H. A. Simon 1954, I. J. Good 1961, 1962).

To give at least a rough idea of how the concept of causality might be introduced into the discussion of various forms of determinism let us turn back to Definition 2.7.1 Suppose that there is a finite number of surrounding conditions $\mathscr{W}_1, \ldots, \mathscr{W}_n$ which define different patterns of behaviour of systems in $\mathscr{X}$. If so, then any change of surrounding conditions of a system $\mathfrak{a}$ in $\mathscr{X}$, say from $\mathscr{W}_i$ to $\mathscr{W}_j$, yields in general a change in the behaviour of $\mathfrak{a}$. Thus either the transition $\mathscr{W}_i \to \mathscr{W}_j$, or the modifications of variables by means of which the surrounding conditions $\mathscr{W}_i$, $\mathscr{W}_j$ are described can be treated as the cause of the event "the change of the pattern of behaviour of $\mathfrak{a}$".

The remarks stated above can easily be recasted in a rigorous and complete manner, in particular the notions involved, especially those of an event and cause, can be defined in a precise fashion. Alas, the analysis of both the causal relations and the correspondence between causality and determinism we can provide going in the direction presented, would be very shallow and consequently disappointing.

The point is that by its very nature the non-statistical concept of causality reduces to the constant succession in time of certain events and thus in fact is redundant (this is exactly Hume's conclusion from his famous analysis of the concept of causality). The statistical concept of causality which Suppes characterizes informally as follows:

> "... one event is the cause of another if the appearance of the first event is followed with a high probability by the appearance of the second, and there is no third event that we can use to factor out the probability relationship between the first and second events".

cannot be subjected to destructive criticism analogous to that provided by Hume.

Our decision to restrict the considerations of regularities to non-statistical ones made at the beginning of this chapter undermines then essentially any possibility of deeper discussion of relationships which causality and determinism bear to each other.

III. EMPIRICAL THEORIES

3.1. AXIOMATIC VERSUS SET-THEORETICAL WAYS OF DEFINING THEORIES

Although in what follows we shall usually assume that each theory possesses a language characteristic of it, it is worth realizing that all such languages can be viewed as fragments of the language of set theory enlarged with additional symbols (terms), when necessary. The latter plays then the role of a universal language of science. Since there are many different systems of set theory (incidentally, the differences among them concern rather sophisticated matters), let us assume that throughout this chapter by set theory we shall mean Zermelo-Fraenkel's version of it, *ZF* for short. The reader may find an exposition of *ZF* in numerous textbooks and monographs. I would especially recommend *An Outline of Mathematical Logic* by A. Grzegorczyk (1974), since the considerations carried out in that book are complementary with respect to many points we are going to study in this chapter.

Let me recall that *ZF* involves only two primitive symbols, Z and $\in$. The latter is a familiar one. Z is a unary predicate symbol; the formula $Z(x)$ reads: *x is a set.* All variables which appear in the formulas of *ZF* are of the same syntactical category, i.e. they can be applied interchangeably in all contexts, although the notation becomes much more legible when variables of a special shape are invented to represent some special sets in which we might be interested, e.g.: relations, functions, power sets, structures of different sorts such as say, topological spaces of Boolean algebras. It is customary, for instance, to use the symbols $R_1, R_2, \ldots$ as variables representing relations (i.e. sets of n-tuples). The procedure by means of which the relation variables can be formally introduced into the language of ZF is simple. To start with one has to define a predicate, say *Rel*, to be able to utter sentences to the effect that x is a relation, $Rel(x)$. The convention should be adopted that any variable R_i must not be applied unless it is explicitly stated, or at least implicitly understood, that $Rel(R_i)$ holds true. In the same way one may introduce special symbol variables for any category of sets whatsoever thus differentiating variables of ZF according to the purpose they are supposed to serve.

An important thing about ZF to remember is that some of the symbols directly involved or definable in ZF do not denote sets. In particular neither Z nor $\in$ are sets, in the sense that the following can be proved in ZF:

(1) There is no z such that: (i) $Z(z)$ and (ii) for all x, $x \in z$ if and only if $Z(x)$,

(2) There is no E such that: (i) $Z(E)$ and (ii) for all x, y $x \in y$ if and only if the pair $(x, y) \in E$.

The predicate *Rel* does not stand for a set, either.

Assertions (1) and (2) seem to be too much involved and one might suggest to replace them by: $\neg Z(Z)$ and $\neg Z(\in)$ respectively. Observe however that neither $Z(Z)$ nor $Z(\in)$ are well-formed formulas of ZF. Although the definition of a well-formed formula of ZF is rather simple we shall not quote it here. For this and other technicalities the reader is advised to consult a suitable text-book, e.g. the one by A. Grzegorczyk mentioned above.

There are two main ways in which the content of a theory Θ can be defined. One may select a certain set of sentences A_Θ to play the role of axioms for Θ, and thus to define

$$\Theta = \{\alpha : A_\Theta \vdash \alpha\}.$$

The other possibility consists in defining a set of interpretations (model structures) M_Θ for the language of Θ and postulating that

$$\Theta = Th(M_\Theta).$$

The latter method of characterizing a theory will be referred to as *set-theoretical* (cf. P. Suppes, 1970a).

If the set of model-structures M_{Θ} is defined, its elements will be referred to as *Θ-structures* or sometimes as *permissible interpretations for* $\mathscr{L}_{\Theta}$.

It is obvious that if Θ is defined in an axiomatic way, one need not worry about defining Θ-structures, and vice versa if Θ-structures are defined one need not be interested in setting up a suitable axiomatics for Θ. On the other hand when axioms for Θ are fixed, they impose some condition on how Θ-structure may be conceived, and similarly a definition of Θ-structure at least suggests an axiomatics for $\mathscr{L}_{\Theta}$.

The main objective of this section is to analyze in more detail the relations between axiomatic and set-theoretical methods of characterizing theories. The discussion will be illustrated by two examples: *Arithmetic of Natural Numbers, AR*, and *theory of consequence, CT*. While *AR* is a purely formal theory, *CT* can be treated as at least semi-factual if not just factual. It may be viewed as a meta-theory which refers to languages actually applied and consequence operations characteristic of them. One may try to interpret *AR* as a factual theory as well, but, I am afraid, this cannot be done in such a straightforward way as in the case of *CT*. On the other hand, however, it is rather obvious that the concept of a natural number is of empirical origin.

The relative simplicity of both *AR* and *CT* makes it possible to state axioms for the two theories in the language of *ZF* enlarged only with the specific terms of the theory in question. Thus for instance in the case of *AR* one has to add to the vocabulary of *ZF* three new primitive symbols: $\mathscr{N}, 0, S$ (which stand for the *set of natural numbers, zero*, and the *successor operation*, respectively). In the case of *CT* two new symbols are needed, $\mathscr{L}$ and *Cn* (the former will be assumed to denote a language and the latter a consequence operation defined in that language).

Let us quote the axioms for *AR* and those for *CT*. The former ones will be exactly the same as one may find in A. Grzegorczyk (1974, p. 213).

AXIOMS FOR *AR*

(AR1) $0 \in \mathcal{N}$,
(AR2) $x \in \mathcal{N} \rightarrow S(x) \in \mathcal{N}$,
(AR3) $x \in \mathcal{N} \rightarrow \neg(0 = S(x))$,
(AR4) $S(x) = S(y) \rightarrow x = y$,
(AR5) $(x \in \mathcal{N} \wedge \neg(x = 0)) \rightarrow \exists y(y \in \mathcal{N} \wedge x = S(y))$,
(AR6) $Z(\mathcal{N})$,
(AR7) $\forall z(Z(z) \wedge 0 \in z \wedge \forall u(u \in z \rightarrow S(u) \in z)) \rightarrow$
$\rightarrow \forall x(x \in \mathcal{N} \rightarrow x \in z)$.

Observe that by setting up axioms for *AR* we have defined an enlargement of *ZF*, denote it as ZF_{AR}, which involves as its theorems all sentences provable from axioms for *ZF* and *AR*. Clearly one may simply identify *AR* and ZF_{AR} but it is more natural to regard *AR* as a proper subtheory of ZF_{AR}. The issue will be discussed later in greater detail.

It is perhaps worth-while noticing that in turn ZF_{AR} may serve as a theory underlying another one in the same way as *ZF* underlies *AR*. In this way a hierarchy of theories can be formed, the specific terms of the theories more fundamental in that hierarchy being auxiliary terms of theories founded on them, in the same way as the specific terms of *ZF* are auxiliary for *AR*.

The axiom system for *CT* we are going to present now will be a slightly modified version of that proposed by A. Tarski (1956, pp. 63, 64).

AXIOMS FOR *CT*

(CT1) $Z(\mathcal{L})$ and $\mathcal{L} \neq \varnothing$,
(CT2) *For all* $X \subseteq \mathcal{L}$, $X \subseteq Cn(X) \subseteq \mathcal{L}$,
(CT3) *For all* $X \subseteq \mathcal{L}$, $Cn(Cn(X)) = Cn(X)$,
(CT4) *For all* $X, Y \subseteq \mathcal{L}$, *If* $X \subseteq Y$, *then* $Cn(X) \subseteq Cn(Y)$,
(CT5) *For all* $\alpha \in \mathcal{L}$ *and for all* $X \subseteq \mathcal{L}$, *If* $\alpha \in Cn(X)$, *then for some finite* $X_f \subseteq X$, $\alpha \in Cn(X_f)$.

Tarski did not demand $\mathcal{L}$ to be non-empty: instead, he required it to be denumerable. Another slight departure from Tarski's original version of axiomatics for *CT* consists in including (CT4) to the axioms. One may easily prove with the help of (CT5) that (CT4) is not independent and thus, in fact, redundant. Sometimes, however, Axiom (CT5) is dropped out and in that case it becomes necessary to have (CT4) among the remaining ones.

By setting up the axioms for a theory Θ one imposes certain conditions to be satisfied by the specific concepts of the theory, thus defining, although indirectly, the class of structures which may be regarded to be permissible interpretations for the specific concepts of the theory. For instance, in the case of *CT*, the acceptance of Axioms (CT1)-(CT5) implies that we expect both $\mathscr{L}$ and *Cn* to satisfy these axioms. The notion of Θ-structure understood in the way described becomes indiscernible from that of a model for Θ.

The last remark calls for some explanation. If one wants to proceed in accordance with what was said in Section 1.9, in order to define a model for *CT* he has to define the language $\mathscr{L}_{CT}$ for *CT*, then the set of standard interpretations $Int(\mathscr{L}_{CT})$ for that language, and finally he might define a model for *CT* to be an interpretation $I \in Int(\mathscr{L}_{CT})$ such that $CT \subseteq Val(I)$. The plan to follow seems to be fairly complicated and thus discouraging. But, fortunately enough, there is another way to achieve the aim.

Actually, we are interested in defining not a model for *CT* but rather a partial model for that theory (a reduct of a model determined by the specific terms of the theory). It is clear that a *partial model* for *CT* should be a structure of the form $(\mathscr{L}^I, Cn^I)$ such that $\mathscr{L}^I$ is a set and Cn^I a unary operation on the set $2^{\mathscr{L}^I}$ of subsets of $\mathscr{L}^I$. It is also obvious that it should be a structure whose constituents $\mathscr{L}^I$, Cn^I satisfy such conditions as Axioms for *CT* imposed on $\mathscr{L}$, *Cn*, respectively.

Note now that we can easily repeat the conditions stated above in the language of *ZF*. If we do not mind using $\mathscr{L}$ and *Cn* as variables (though in axiomatics for *CT* they appear as constant symbols) we may word the definition of a partial model for *CT* as follows.

DEFINITION 3.1.1. *A structure* $(\mathscr{L}, Cn)$ *is a* partial model for *CT* *if and only if* $\mathscr{L}$ *is a non-empty set,* $Cn: 2^{\mathscr{L}} \to 2^{\mathscr{L}}$, *and Axioms* (CT2)-(CT5) *are satisfied.*

Clearly, instead of saying that (CT2)-(CT5) are satisfied one may write these conditions explicitly. Note that (CT1) has already been stated in that way.

Although the concept of a partial model for AR can be defined in a fully analogous way, one must realize that the definition obtained in that way would not be the best one. The reason for that is that in the axioms for AR the notion of a set intervenes much more essentially than in CT, where the sets considered are restricted, due to suitable explicit conditions imposed on the range of variables $X, Y, \ldots, \alpha, \beta, \ldots$ appearing in the axioms, to those being subsets of $\mathscr{L}$. As for AR, nothing compels us to treat variables x, y, z as running over $\mathscr{N}$. Moreover, such a restriction seems to be unjustified. It turns out then that the notion of a model for AR is more involved than that for CT, since we must somehow settle how the set-theoretical concepts involved in AR are to be defined.

A possible way of solving the problem could be the one we have already mentioned while discussing the notion of a partial model for CT. First one should define the notion of a possible interpretation for the language of ZF_{AR}, i.e. for all the symbols $\in, Z, \mathscr{N}, 0, S$, and then in each of such interpretations separate the part relevant to $\mathscr{N}, 0, S$ (the $\mathscr{N}, 0, S$-reduct of the interpretation). An attempt to implement this program makes it necessary to transfer the discussion to a metalanguage for ZF. It certainly can be done, but we want to treat ZF as an ultimate foundation of any theory whatsoever, and consequently avoid any considerations in which ZF itself is subjected to examinations.

Seemingly we have complicated our problem by banning a certain approach, but there is another and considerably simpler way to cope with it.

Given an arbitrary set U, define Z_U and $\in_U$ as follows:

(Z_U) $\quad Z_U(x)$ if and only if $Z(x)$ and $x \in U$,

($\in_U$) $\quad x \in_U y$ if and only if $x \in y$ and $x, y \in U$.

With the help of the symbols Z_U and $\in_U$ we may easily define the notion of a partial model for AR as follows.

DEFINITION 3.1.2. (cf. A. Grzegorczyk, 1974, p. 348). *A structure* $(\mathcal{N}, 0, S)$ *will be said to be a* (partial) model for *AR if and only if for some infinite set U such that* $\mathcal{N} \subseteq U$ *and with every set X which U includes it also includes* 2^X, *the following conditions hold true*:

i. $0 \in_U \mathcal{N}$,
ii. $x \in_U \mathcal{N} \to S(x) \in_U \mathcal{N}$,
iii. $x \in_U \mathcal{N} \to \neg(0 = S(x))$,
iv. $S(x) = S(y) \to x = y$,
v. $(x \in_U \mathcal{N} \wedge \neg(x = 0)) \to \exists y (y \in_U \mathcal{N} \wedge x = S(y))$,
vi. $Z_U(\mathcal{N})$,
vii. $\forall z (Z_U(z) \wedge 0 \in_U z \forall u (u \in_U z \to S(u) \in_U z)) \to$
$\to \forall x (x \in_U \mathcal{N} \to x \in_U z)$.

Let us now examine the possibility to alternate the procedures described, i.e. let us examine the possibility to define the permissible interpretations for the language of a theory, still before a set of axioms for that theory has been set up.

When trying to set up axioms for a theory Θ, one usually must observe certain linguistic constraints: the axioms should not involve terms which are not supposed to appear in the theorems of the theory. The list of the terms which are not welcome in the axioms need not be explicitly given, but the existence of such a list seems to be characteristic of any attempt to set up a theory in an axiomatic manner. One certainly might introduce, say, the concept of proof, or that of a rule of inference to *CT*, but the resulting theory would not be that intended by A. Tarski who wanted to study just the consequence operation and nothing else. Similarly one may introduce the concept of cardinality of a set into *AR*, but again, the resulting theory would be a supertheory of *AR*, essentially stronger than the one we have been accustomed to study under the heading "Arithmetic of Natural Numbers".

The constraints of this sort do not intervene when one wants to define models for the specific concepts the theory involves, and this is perhaps the main reason why it might be desirable to discuss the concept of a Θ-structure without presupposing anything about how it might refer to the axioms for Θ. Let us discuss the concept of *AR*-structure and that of *CT*-structure.

The intuitive idea of a natural number can be accounted for by means of so-called *genetic construction* of natural numbers, which we are going to present in a sketchy way. For a more detailed discussion cf. A. Grzegorczyk (1974).

To begin with let us remind that two sets A, B are said to be *equinumerous*, $A \sim B$, if there is a one-to-one mapping of one of these sets onto the other. A finite set is the one which is not equinumerous with any of its proper subsets; perhaps the adequacy of this definition is not obvious at the first glance, but the issue is too involved technically and too special to be discussed here. The family of finite subsets of a set A will be denoted as $Fin(A)$.

Let U be an infinite set. Define the symbols $card_U$, N_U and S_U as follows:

(N1) $card_U(A) = \{B \in Fin(U)\colon B \sim A$ *for any* $A \in Fin(U)$,

(N2) $N_U = \{card_U(A)\colon A \in Fin(U)\}$,

(N3) $S_U(card_U(A)) = card_U(A \cup \{a\})$ *for any* $a \in U - A$.

The last definition should be preceded by a proof of existence of exactly one function satisfying the conditions imposed on S_U. But again because of the technical nature of the problem we shall not undertake it.

Now, we are in a position to formulate the definition in which we are interested.

DEFINITION 3.1.3. *A structure* $(N, 0, S)$ *will be said to be an* *AR*-structure, *if and only if for some infinite set* U *the following conditions are satisfied*:

Axiom A1. $N = N_U$,

Axiom A2. $0 = card_U(\varnothing)$,

Axiom A3. $S = S_U$.

In turn let us examine the concepts involved in the axioms for *CT* from the set-theoretical point of view. Note first that on this level of generality which is accepted in *CT*, any non-empty set of object may be treated as a language. Obviously one may call these object signs and figure out that each of such signs conveys certain information, i.e. it functions as a sentence. It is rather clear that any object whatsoever can be used as a sign under suitable circumstances. Axiom (CT1) provides then indeed a complete characteristic of the language understood in such a general fashion.

Now, consider the consequence operation. The notion of a consequence is strongly related to that of a proof. To say that a sentence α is a consequence of X amounts to saying that α is provable from X. In turn the notion of a proof is, as we remember (cf. Section 1.10), definable in terms of rules of inference. A proof is a sequence of sentences such that each element in the sequence is either a premise (i.e. an element of the set X from which we try to derive the hypothesis being proved) or is derivable from the earlier one in the sequence by the rules of inference accepted.

To match this level of generality which is characteristic of Tarski's axiomatics for CT, one should define a rule of inference to be a relation of the form $R(X, \alpha)$ where $X \subseteq L$ and $\alpha \in \mathscr{L}$. If α bears the relation R to X we say that α is derivable from X in one step.

DEFINITION 3.1.4. *A structure* $(\mathscr{L}, Cn)$ *will be said to be a CT-structure if and only if the following axioms are satisfied*:

Axiom C1. $\mathscr{L}$ *is a non-empty set.*

Axiom C2. Cn *is an operation defined on subsets of* $\mathscr{L}$, *i.e.* $Cn: 2^{\mathscr{L}} \to 2^{\mathscr{L}}$.

Axiom C3. *There is a set* $\mathscr{R}$ *of rules of inference defined on* $\mathscr{L}$ *such that for each* $X \subseteq \mathscr{L}$ *and for each* $\alpha \in \mathscr{L}, \alpha \in Cn(X)$ *if and only if there is a proof of* α *from* X *by means of rules in* $\mathscr{R}$.

The reason why Definition 3.1.4. seems to provide a deeper insight into the nature of the consequence operation is obvious. One may not doubt the adequacy of Axioms CT1 – CT5 but still may wonder whether they are complete, i.e. whether they do not have to be complemented by some additional ones. The issue cannot be examined in a rigorous manner unless a mathematical model for both the concept of a language and that of a consequence operation is accepted, i.e. unless the notion of CT-structure is defined. Similar remarks apply to AR.

Before we compare the notion of an AR-structure with that of a partial model for AR, and similarly the notion of CT-structure with that of a partial model for CT, let us explain how the definition of a Θ-structure can be exploited in order to define the content of Θ.

In fact the thing is simple. The axioms which serve to define the notion of a Θ-structure may involve some concepts which are desired neither as specific nor as auxiliary concepts of Θ. Still if one has a clear idea of how the language $\mathscr{L}_\Theta$ of Θ should be defined, he may define the content of Θ as the set of all sentences of $\mathscr{L}_\Theta$ provable from the axioms for Θ-structures. We shall discuss this procedure in greater detail in the next section.

One important point however should be brought out. In order to get an immediate idea of what we have in mind consider Definition 3.1.3. The only genuinely new term added to the language of ZF after accepting that definition is "AR-structure". The symbols $\mathscr{N}, 0, S$ which appear in the definition are novel only seemingly! They are just variable symbols of exactly the same sort as any variables of ZF whatsoever, and the definition would not be affected at all if one replaced $\mathscr{N}, 0, S$ by, say, x, y, z respectively.

The reason why we decide to use $\mathscr{N}, 0, S$ rather than x, y, z is, however, transparent. In some contexts we want to apply them as constant symbols, corresponding to the symbols $\mathscr{N}, 0, S$ characterized axiomatically. At that very moment we start treating Axioms A1-A3 as axioms for $\mathscr{N}, 0, S$ not only as conditions by means of which the predicate AR-structure is characterized. In that manner Definition 3.1.1 provides us, although implicitly, with certain alternative system of axioms for AR.

The discussion we have carried out brings to light that theories need not be conceived as sets of theorems. One may identify the theory with the conceptual apparatus characteristic of it: in order to define a theory it is enough to define the way in which the specific concepts of the theory should be interpreted.

In fact each theory plays a two-fold role. It summarizes our knowledge on certain phenomena and at the same time provides a certain conceptualization for them. Therefore when we decide to define a theory Θ by defining the notion of a Θ-structure we have to verify whether the definition proposed conforms to the theorems we are ready to accept as theorems of Θ. On the other hand, when we decide to define the theory by setting up a suitable set of axioms for it, we have to verify whether the interpretations for the specific concepts provided by the (partial) models for the axioms fit the way of conceiving the concepts we consider to be the proper one. Obviously the issue can be discussed in a rigorous manner only exceptionally when both the content and the conceptual apparatus of the theory are separately and rigorously defined.

This is the situation we face in the case of both AR and CT. One may use Definition 3.1.3 in order to examine the adequacy of Axioms (AR1) – (AR7), and vice versa, one may use Axioms (AR1) – (AR7) in order to verify the adequacy of the genetic construction of natural numbers. The same can be said about CT. One may either exploit Definition 3.1.4 in order to verify the adequacy of Axioms (CT1) – (CT5), or conversely he may exploit (CT1) – (CT5) in order to verify the adequacy of Definition 3.1.4. Let us discuss the matter in a concise manner without going into technical details.

Given a theory Θ defined both in axiomatic and set-theoretical manner one may ask two questions:

(Q1) Is the content of the theory in both the cases the same?

(Q2) Does the notion of Θ-structure coincides with that of a partial model for Θ?

In general, question Q1 should be relativized to a suitably selected language. In the same way as ZF_{AR} has been defined, one may define ZF'_{AR}, using Axioms A1 – A3 involved in Definition 3.1.4 instead of (AR1) – (AR7). Thus ZF'_{AR} is the set of all sentences of the language of ZF enlarged with $\mathcal{N}, 0, S$, which are provable from the axioms for ZF and A1 – A3.

If the content of AR is defined to be either ZF_{AR} or ZF'_{AR}

depending on the way in which AR is set up, one may easily verify that these two ways of defining AR do not result in the same theory, i.e. $ZF_{AR} \neq ZF'_{AR}$. Note for instance that Axiom A2 of Definition 3.1.3 entails that

(1) $\emptyset \in 0$,

while Axioms AR1 – AR7 allow to prove neither that $\emptyset \in 0$ nor even that $Z(0)$. For another example, consider the sentence:

(2) *For all* $x, y \in \mathcal{N}$, *either* $x = y$ *or* $x \cap y = \emptyset$.

Again sentence (2) is true when $\mathcal{N}$ is understod in the sense defined by Axiom A1 of Definition 3.1.3, but it can be neither proved nor disproved on the base of Axioms AR1 – AR7.

It would be hardly reasonable to question the completeness of Axioms (AR1)-(AR7) just because they do not allow us to prove (1) and (2). After all they are "pathological" assertions. The question whether the empty set is an element of 0 or whether two natural numbers are sets with a common element are from an intuitive point of view meaningless.

It may be proved that under a certain "reasonable" definition of the language $\mathscr{L}_{AR}$ of AR we have

$$ZF_{AR} \cap \mathscr{L}_{AR} = ZF'_{AR} \cap \mathscr{L}_{AR}.$$

In particular this holds true when $\mathscr{L}_{AR}$ is defined to be an elementary language (see section 3.2). On the other hand we have:

ASSERTION 3.1.5. *Each partial model for AR is isomorphic to an AR-structure. At the same time all AR-structures are partial models for AR.*

(As a matter of fact all partial models for AR are isomorphic, cf. Grzegorczyk, 1974, p. 349, Theorem 40).

One of the reasonable requirements that one may expect to be imposed on the language $\mathscr{L}_\Theta$ of Θ is that the sentences of that language preserve their truth-values under isomorphic transformations of interpretations of $\mathscr{L}_\Theta$, in particular partial models for Θ. Note that neither Assertion (1) nor (2) are sentences of this kind. This accounts for their pathological nature perhaps better than the remarks we made earlier.

The case of CT is much less complicated than that of AR. One may prove that CT is the same theory regardless of whether it is considered to be set up in axiomatic or set-theoretical manner. Thus, in particular, if ZF_{CT} and ZF'_{CT} are defined in the way analogous to that in which ZF_{AR} and ZF'_{AR} have been defined, then

$$ZF_{CT} = ZF'_{CT}.$$

At the same time each partial model for CT is a CT-structure and vice-versa.

3.2. THEORIES AS DEDUCTIVE SYSTEMS

Let us examine closer the problems to which an attempt to define the language and then the content of a theory may give rise. To begin with let us enlarge our fairly meager list of examples of theories with some new ones.

EXAMPLE 1. In what follows we shall often use the concept of probability. Let us therefore quote the definition of a probability space. The conditions by means of which this concept is characterized serve as axioms for the probability calculus, PC – for short. Probability spaces are thus PC-structures in the terminology

DEFINITION 3.2.1.

A. *A structure* $\mathfrak{P}$ *is a* finitely additive probability space *if and only if* $\mathfrak{P}$ *is a triple* $(S, \mathscr{S}, P)$ *such that the following axioms are satisfied*

Axiom P1 *S is a non-empty set,*

Axiom P2 *$\mathscr{S}$ is an algebra of sets on S,*

Axiom P3 $P: \mathscr{S} \to [0, 1]$,

Axiom P4 $P(S) = 1$.

Axiom P5 *If* $A \cap B = \varnothing$, *then* $P(A \cup B) = P(A) + P(B)$.

B. *Moreover,* $\mathfrak{P}$ *is a* probability space (*without restriction to finite additivity*) *if the following two axioms are also satisfied*:

Axiom P6 $\mathcal{S}$ *is a* σ*-algebra of sets on* S;

Axiom P7 *If* $A_1, A_2, \ldots$ *is a sequence of pairwise disjoint elements of* S *i.e.,* $A_i \cap A_j = \varnothing$ *for* $i \neq j$, *then*

$$P(\bigcup_{i=1}^{\infty} A_i) = \sum_{i=1}^{\infty} P(A_i).$$

In the probability calculus the algebra of sets $\mathcal{S}$ is usually referred to as *algebra* (or *space*) *of events*, and consequently, elements of $\mathcal{S}$ are called *events*. The set S is called a *sample space*, its elements being often referred to as *outcomes* (of some experiment). Given an A in $\mathcal{S}$, $P(A)$ is referred to as the *probability* of A.

Since we have decided to recall specific terminology of the probability calculus, alhough the reader is most likely familiar with it, let us also remind that if $A \cap B = \varnothing$, then the events A and B are said to be *incompatible*, $\varnothing$ is the *impossible event*, while S is the certain event. The event $-A = S - A$ is called the *complement* of the event A, $\bigcup A_i$ (where $i = 1, 2, \ldots$) is called the union, and $\bigcap A_i$ is called the intersection of the events A_i. Thus in the last three cases the set-theoretical and probabilistic terminologies coincide. Evidently $-A$ is the event that A does not occur, $\bigcup A_i$ is the event which consists in the occurrence of at least one of the events A_i, $\bigcap A_i$ is the event which consists in the occurrence of each of the events A_i.

The fact that each Boolean algebra is isomorphic to an algebra of sets (*Stone's Representation Theorem*) allows us to extend in an obvious way the notion of the probability onto Boolean algebras (elements of the Boolean algebra on which a probability measure is defined are then treated as events). This possibility will be exploited later without any additional comments.

EXAMPLE 2. The theory we shall characterize now, again in the set-theoretical manner, will be assumed to be a theory of the linear hierarchy pattern observed in some groups of animals, for instance in a small group of hens which live together for a long time. At present we shall mainly be interested in the formal structure of the theory; a discussion of possible empirical interpretations of the theory will be postponed till Section 3.5. Still a few comments on the intended interpretation of the symbols which will appear in the definition which we are going to state should be of some help.

A *linear hierarchy structure* (*LH-structure*, for short) will be defined to be a structure which involves four constituents denoted as G, D, P, U. G should be conceived as a group of animals, and U as a set of objects for which the individuals in G can run competing among themselves in achieving them. Let us call these objects *targets*. The elements of U may be for instance: pieces of food, mating partners, shelters, direct contact with a protector. D under the intended interpretation stands for domination relation, thus $D(a, b)$ reads: *a dominates b*. Finally, P will be applied in the contexts of the form $P_{X,u}(a, k) = p$, and we shall assume that the intended meaning of such a formula is: the probability of the event that a attains the target u at trial k and in the situation when X is the set of all individuals in G competing for u is p. Obviously we assume here that the situation of competing for the same object by the individuals of the same group of animals can be repeated; each repetition forms a trial. We also assume that at each trial there is only one animal which attains the target.

DEFINITION 3.2.2. *A structure* (G, D, P, U) *will be said to be an* LH-structure (*a* linear hierarchy structure) *if and only if the following axioms are satisfied*:

Axiom D1. *G is a non-empty finite set.*

Axiom D2. *U is a non-empty set.*

Axiom D3. *For all $u \in U$ and for all non-empty subsets $X \subseteq G$, $P_{X,u}$ is a probability measure defined on the σ-algebra of subsets of the set of all sequences of elements of G.*

Axiom D4. *D is a binary relation on G such that for all $a, b \in G$, $D(a, b)$ if and only if for all $u \in U$ and for all $X \subseteq G$, and for all trials k, $P_{X,u}(a, k) \geqslant P_{X,u}(b, k)$, provided that $a \in X$.*

Axiom D5. *For all* $a, b \in G$, $D(a, b)$ *or* $D(b, a)$.

Axiom D6. *If* $a \notin X$, *then for all* u *and for all trials* k, $P_{X,u}(a, k) = 0$.

Axiom D7. *For all* $a \in G, X \subseteq G, u \in U$, *and for all trials* k, $P_{X,u}(a, k) \approx 1$ *or* $P_{X,u}(a, k) \approx 0$.

Different hierarchy patterns observed in animal groups are usually described in the conversational language improved by enriching it with some technical terms. It is perhaps worth realizing that even so simple, from the mathematical standpoint, a theory as that of linear hierarchy, *LH*, may be relatively easily presented in a rigorous form, i.e. as an applied mathematical theory. This has motivated our selection of *LH* as an example to illustrate the later discussion, although the example is artificial in the sense that nobody has ever proposed to formalize *LH* in the way we did.

EXAMPLE 3. The last theory we want to define is that of particle mechanics, *PM* for short. Let us first state the definition of *PM*-structure and only then comment briefly on it, in order to explain some of the symbols and concepts it involves.

DEFINITION 3.2.3. *A structure* $(\mathscr{C}^s, \mathscr{K}, T, q, L)$ is a *PM*-structure *if and only if the following axioms are satisfied*:

Axiom M1. $\mathscr{C}^s$ *is the s-dimensional configurational space.*

Axiom M2. K *is a frame of reference for* $\mathscr{C}^s$.

Axiom M3. T *is an interval of real numbers* (*time interval*).

Axiom M4. q *is a curve in* $\mathscr{C}^s \times Re$, *twice differentiable with respect to the* (*time*) *variable running over* T.

Axiom M5. *The function* L (*called the* Lagrangian function) *is of the form* $L(q, \dot{q}, t)$, *i.e. the values of* L *are definable in terms of the values of* q, $\dot{q}$, *and* t, *where* $\dot{q}$ *is the first derivative of* q *relative to time.*

Axiom M6. *For each* t *in* T, *the values of the coordinates* $q_i(t)$, $i = 1, \ldots, s$, *of* q *are uniquely determined by a system of equations*

$$\frac{d}{dt}\frac{L}{\dot{q}i} - \frac{L}{qi} = 0,$$

provided that for some $t_0 \in T$ *the values of both* $q_i(t_0)$, and $\dot{q}_i(t_0)$ are given for all $i = 1, \ldots, s$.

The definition of a PM-structure we have proposed is not complete; it might be supplemented by some additional conditions (axioms) imposed on the specific concepts involved, for instance one may demand L to be an additive function in a very special physical sense (If L_A, L_B are the Lagrangians of two systems such that there is no interaction betweem them, then $L_{A+B} = L_A + L_B$, where L_{A+B} is the Lagrangian of the physical union $A+B$ of the systems A, B). We are not interested, however, in full adequacy of our considerations from the point of view of physics, what we need is just a general idea of how empirical theories can be defined in a rigorous way, and how such definitions may work.

Let us begin our comments on Definition 3.2.3. with a short explanation of how a curve in the s-dimensional configurational space $\mathscr{C}^s$ models the movements of empirical bodies (particles). After all, particle mechanics is a theory of physical systems located in the usual physical space.

The notion of a particle can be characterized as follows:

> "The name denotes a body whose size can be neglected, when describing its motion. The possibility to neglect size of a body depends of course on specific conditions of a given problem. Thus, when examining for instance the revolution of planets round the sun, one may consider the planets to be particles, whereas when describing their twenty-four-hours' rotation, the procedure is inadmissible."

(cf. L. D. Landau and E. M. Lifshic, 1969.)

This definition brings an attempt to explain what objects can be referred to as particles from the point of view of physics. From the mathematical viewpoint the particle may be treated as an arbitrary point of the three-dimensional Euclidean space $\mathscr{E}^3$ (we assume that the three-dimensional Euclidean space is a model of the physical one).

To define location of a particle in a space we need a system of coordinates. From the physical viewpoint a system of coordinates is some physical body. However, a purely mathematical definition of the notion is also possible. For instance, one may assume (although the definition obtained in this way is a bit too narrow) that a system of coordinates is the basis of the space $\mathscr{E}^3$ transforming in time so that neither angles between vectors nor lengths of the vectors change. The assumption of transformation in time allows to treat systems of reference as being in motion with respect to each other and thus also with respect to the "absolute" physical space. In accordance with the fundamental principles of non-relativistic mechanics we assume that the set of time-instants is isomorphic to the set of real numbers, and the function which settles isomorphism between the two sets is established once and for all.

Suppose we deal with a system of N particles. It may turn out that the positions the particles can take are not quite arbitrary, and thus it is not necessary to know all the $3N$ coordinates to define uniquely these positions. The least number of space coordinates which has to be known in order to define positions of all particles of the system is said to be a number of the *degrees of freedom* of the system.

To describe behaviour of a system $\mathfrak{A}$ with s degrees of freedom it suffices to concentrate on s suitably chosen quantities $q_1, q_2, \ldots, q_s$ and trace changes of these quantities in time. From the mathematical point of view it means that instead of examining the behaviour of N points in the three-dimensional space we pass on to examining the behaviour of only one point in the s-dimensional space, called a *configurational space.* The quantities $q_1, \ldots, q_s$ are coordinates of the configurational space, or *generalized coordinates.* Experience shows that the values the quantities $q_1, \ldots, q_s$ and their derivatives take at an arbitrary time-instant t_0 determine uniquely the state of the system $\mathfrak{A}$ at any other time-instant t.

All the examples we have produced in this section are examples of the theories which, regardless of the degree of their sophistication, involve a pretty large number of auxiliary mathematical concepts. Investigations within probability theory cannot be effectively carried out without applying methods (and thus also concepts) of mathematical analysis (differential and integral calculi). Theory of linear hierarchy presupposes some results of probability theory. Particle mechanics is based on vector analysis.

The auxiliary concepts the theory involves belong to the vocabulary of the language of the theory on a pair with the specific ones. Unfortunately, quite often to produce an exhaustive list of them is not an easy task. Even if a theory exploits a small amount of auxiliary concepts, nobody can ever predict which of the mathematical methods will turn out useful in the future in dealing with the problems of the theory. It is undesirable then to limit in any way whatsoever the language of the theory; no sound mathematical concepts and no sound mathematical methods must be ruled out.

It should perhaps be made clear then that to define the language of a theory having in view the needs of methodology of science and that of the theory itself are two different things. The more rigorous the methods applied in the course of methodological inquiry are, the more rigorous the definition of the theory subjected to examinations must be. In particular, investigations concerning such notions as: inconsistence of a theory, definability of some primitive concept in terms of others, soundness of a theory with respect to a given application etc, etc, cannot be carried out in a fully precise way unless the language of the theory is defined. The methodological assumptions concerning the properties of the theory examined are by no means principles to be observed by the people who develop the theory. One should not expect that if the methodologist decides to study, say, particle mechanics under the assumption that the language of *PM* is based on a certain definite and fixed list of terms, the physicist who undertakes the problems of particle mechanics must follow his assumptions. As a matter of fact, even in methodological investigations concerning *PM* the language of *PM* may be defined in many different ways. The choice of the definition is dictated both by the problem posed and by the properties of the real-life particle mechanics, i.e. particle mechanics meant to be a social phenomenon: a product of the scientists' activity.

The simplest way in which the language of a theory can be defined consists in listing the symbols (terms) which will be allowed to appear in the sentences of the language in question and to assume that each well-formed and closed formula which involves only the singled out symbols is a sentence of that language. We may be interested however, in defining the set of sentences in a more restrictive way, by ruling out certain contexts (i.e. concatenations of symbols). Let us illustrate the last remark with a few examples.

Define the set $\mathcal{L}_{AR}$ of sentences of ZF_{AR} as the set of sentences which apart from variables, logical symbols, and the symbols $\mathcal{N}, 0, S$, involve only some symbols of operations on $\mathcal{N}$ and subsets of $\mathcal{N}$, defined in ZF_{AR}. Thus the symbol Z does not appear in the sentences of $\mathcal{L}_{AR}$. Moreover assume that the only contexts in which the symbol $\in$ may appear are formulas of the form $x \in K$, where K is a name for a subset of $\mathcal{N}$. $\mathcal{L}_{AR}$ does not involve then sentences such as $\mathcal{N} \in 2^{\mathcal{N}}$ or $Z(S(\emptyset))$ which are well formed formulas of ZF_{AR}.

The following two features of $\mathcal{L}_{AR}$ are worth mentioning.

(i) If there is only a finite number of defined symbols which appear in the sentences of $\mathcal{L}_{AR}$, $\mathcal{L}_{AR}$ is defined in an effective way (i.e. there is a procedure which allows us to decide about any sentence of Z_{AR} whether it is an element of $\mathcal{L}_{AR}$ or not).

(ii) The role of $\in$ in the sentences of $\mathcal{L}_{AR}$ is a rudimentary one. In fact, instead of $x \in K$, one may write $K(x)$, which brings to light that all sentences in $\mathcal{L}_{AR}$ may be easily translated onto the elementary language (the language of the first order predicate calculus). The sentences in $\mathcal{L}_{AR}$ may be then called the *elementary* ones.

In view of (i) in order to transform $\mathcal{L}_{AR}$ into a formalized language it is enough to define the notion of derivability $\vdash_{AR}$ pertinent to $\mathcal{L}_{AR}$. Let us discuss the matter in a fully general way referring to $\mathcal{L}_{AR}$ only for illustration.

Given any extension ZF_θ of ZF denote by $\vdash$ derivability in ZF_θ determined by the logical axioms and rules, axioms of ZF, and definitions accepted in ZF_θ. More precisely, $X \vdash \alpha$ is assumed to be valid if and only if there is a sequence

$$\alpha_1, \alpha_2, \ldots, \alpha_k$$

(a *proof of* α *from* X) such that $\alpha = \alpha_k$, and for each α_i

($i = 1, \dots, k$) in the sequence at least one of the following conditions holds true:

(P1) α_i is in X.
(P2) α_i is of the form of a logical axiom.
(P3) α_i is an axiom of ZF.
(P4) α_i is a definition accepted in ZF_Θ.
(P5) α_i results from $\alpha_1, \dots, \alpha_{i-1}$ by a logical rule of inference (we assume that both the set of logical axioms and that of logical rules of inference are defined).

A natural solution to the problem of defining the derivability relation pertinent to $\mathscr{L}_\Theta$ seems to consist in postulating $X \vdash_\Theta \alpha$ to hold true if there is a proof of α from X constructed thoroughly of the elements of $\mathscr{L}_\Theta$. Note however that neither axioms of ZF nor definitions accepted in ZF_Θ need to be elements of $\mathscr{L}_\Theta$. For instance most of the axioms of ZF are not elements of $\mathscr{L}_{AR}$. Since we may not give up referring to set theoretical properties of the concepts involved in $\mathscr{L}_\Theta$, we must reconsider the proposed definition of $\vdash_\Theta$.

Given any $X \subseteq \mathscr{L}_\Theta$ and any $\alpha \in \mathscr{L}_\Theta$ we shall define $X \vdash_\Theta \alpha$ to hold true if and only if $X \vdash \alpha$. In that way the existence of a proof of α from X, comprising only sentences from $\mathscr{L}_\Theta$, is not a neccessary condition for α to bear the relation $\vdash_\Theta$ to X. Let me point out, however, that the following is trivially valid.

Assertion 3.2.4. *Let ZF_Θ and $\mathscr{L}_\Theta$ be as described above and let $X \subseteq \mathscr{L}_\Theta$, $\alpha \in \mathscr{L}_\Theta$. Then $X \vdash_\Theta \alpha$ if and only if there is a proof*

$$\alpha_1, \dots, \alpha_k$$

of α from $X \cup \{\beta \in L_\Theta : \vdash \beta\}$ such that all elements α_i of the proof are in $\mathscr{L}_\Theta$.

Proof. If $X \vdash \alpha$, then in view of the finitary nature of the derivability relation, $\alpha_1, \dots, \alpha_n \vdash \alpha$ for some $\alpha_1, \dots, \alpha_n$ in X. By Deduction Theorem we conclude that

$$\alpha_1 \wedge \dots \wedge \alpha_n \Rightarrow \alpha$$

is a mathematical identity and thus an element of the set $\{\alpha \in \mathscr{L}_\Theta : \vdash \alpha\}$. The sequence

$$\alpha_1, \ldots, \alpha_n, \alpha_1 \wedge \ldots \wedge \alpha_n \Rightarrow \alpha, \alpha$$

forms the proof of α of the needed sort.

In what follows we shall refer to the elements of

$$\{\alpha \in \mathscr{L}_\Theta : \vdash \alpha\}$$

as to $\mathscr{L}_\Theta$*-identities*. Observe that in general there is no guarantee that the set of $\mathscr{L}_\Theta$*-identities* is axiomatizable, it need not be such.

The main advantage of the way in which we have decided to define $\vdash_\Theta$ lies in its simplicity. Besides, it yields a well defined notion of the derivability $\vdash_\Theta$ regardless of how the set $\mathscr{L}_\Theta$ has been formed and what sentences it includes. Note also that the definition eventually accepted allows us to drop the subscript Θ and write simply $\vdash$ instead of $\vdash_\Theta$. This not only facilitates the notation but also calls the attention to the fact that the notion of derivability will be applied in a certain unified manner, loosely speaking the same in which that notion is applied in mathematics.

Let us produce another example to illustrate the present considerations concerning the notion of the language of a theory.

Define $\mathscr{L}_{CT}$ to be the set of all sentences of ZF_{CT} which can be formed by means of: variables, logical symbols, the specific terms L, Cn of CT, and the following set-theoretical symbols: $\subseteq, \cup, \cap, -, \varnothing$ and Fin, where $Fin(X)$ reads: X is a finite set. We assume that the definition of Fin is given in ZF_{CT}. Observe that just like $\mathscr{L}_{AR}$, $\mathscr{L}_{CT}$ can be treated as an elementary language. At the same time it is easily seen that all axioms for CT can be easily casted in $\mathscr{L}_{CT}$. Indeed, define $1(X)$ as follows:

(1) $1(X)$ *if and only if* $X \neq \varnothing$ *and for all* $Y \subset X, Y = \varnothing$.

Now Axiom (CT4), the only one which does not belong directly to $\mathscr{L}_{CT}$, can be rephrased in the following way:

(CT4′) *If* $1(X)$ *and* $X \subseteq Cn(Y)$, *then for some* $Z \subseteq Y$, $Fin(Z)$ *and* $X \subseteq Cn(Z)$.

There is one thing which must be realized here. The language $\mathscr{L}_{CT}$ may indeed be treated as an elementary one: the variables of that language run over a set of objects of the same set-theoretical type (subsets of $\mathscr{L}$) and the constant symbols are symbols of operations on those objects and relations between them. But this fact does not imply that the logic underlying $\mathscr{L}_{CT}$ is elementary as well! The symbols like 1, $\emptyset$, *Fin* have their set-theoretical meaning which intervenes in all arguments involving these symbols. In particular then the set of mathematical identities $\{\alpha \in \mathscr{L}_{CT}: \vdash \alpha\}$ of $\mathscr{L}_{CT}$ includes, apart from logical tautologies, some set-theoretical theorems (e.g. If $1(X)$, then $\emptyset \neq X$, $Fin(\emptyset)$) and the derivability relation $\vdash_{CT}$ is stronger than that defined by the rules of the first order logic only.

Consequently, the set of standard interpretations for $\mathscr{L}_{CT}$ should be defined as comprising only some of the structures which would provide interpretation for $\mathscr{L}_{CT}$ if $\mathscr{L}_{CT}$ were defined as based on elementary logic. Incidentally the notion of a standard interpretation for $\mathscr{L}_{CT}$ can easily be defined in a rigorous way in *ZF*.

DEFINITION 3.2.5. *A structure I is a* standard model-structure for $\mathscr{L}_{CT}$ *if and only if I is a structure of the form* $(2^{\mathscr{L}}, \mathscr{L}, Cn, \subseteq, \cup, \cap, -, Fin, 1, \emptyset)$, where:

(a) $\mathscr{L}$ *is a non-empty set,*
(b) $Cn: 2^{\mathscr{L}} \to 2^{\mathscr{L}}$,
(c) $\subseteq$ *is inclusion relation restricted to* $2^{\mathscr{L}}$,
(d) $\cup$, $\cap$, $-$ *are union, meet and relative complement operations restricted to* $2^{\mathscr{L}}$,
(e) *Fin is the class of finite subsets of* $\mathscr{L}$,
(f) 1 *is the class of unit subsets of* $\mathscr{L}$, *and*
(g) $\emptyset$ *is the empty set.*

Were $\mathscr{L}_{CT}$ assumed to be based on elementary logic, its permitted interpretations would be all structures similar to that defined above, i.e. any binary relation on $2^{\mathscr{L}}$ could be selected to be an interpretation for $\subseteq$, any binary operation on $2^{\mathscr{L}}$ could be selected to be an interpretation for $\cup$, etc...

Clearly it is enough to point out only two constituents of

a standard interpretation for $\mathscr{L}_{CT}$, namely $\mathscr{L}$ and *Cn*, to determine the whole interpretation. This allows us to treat structures of the form $(\mathscr{L}, Cn)$, with $\mathscr{L}$ and *Cn* satisfying conditions (a) and (b) of Definition 3.2.5, as standard model-structures for $\mathscr{L}_{CT}$ and some of them as partial models for *CT*.

The comments on the logic underlying $\mathscr{L}_{CT}$ and the set of permissible interpretations for that language may suggest that the fact that a language $\mathscr{L}_\Theta$ of a theory Θ was defined to be elementary (or translatable onto elementary) is of minor significance, and in particular does not open the possibility to apply the wide bulk of methods and results concerning elementary theories to the theory Θ. But the situation is not so bad.

Note that in view of Assertion 3.2.4 any theory Θ whose axioms have been stated in an elementary language $\mathscr{L}_\Theta$ coincides with the theory Θ' whose content is defined to be the set of all logical consequences (sentences derivable by logical rules only) of the union of the set of the axioms for Θ and the set of mathematical identities $\{\alpha \in \mathscr{L}_\Theta : \vdash \alpha\}$ of $\mathscr{L}_\Theta$. It should be stressed, however, that when treated as elementary, theory Θ need not be either finitely or recursively axiomatizable.

Although the task of transforming a theory into an elementary one (not only from the linguistic but also from the deductive and semantic point of view) can be formidable, it is worth realizing that even relatively complicated theories can be presented in that form. For instance, as it was shown by R. Montague (1962), this can be done in the case of particle mechanics (at least for some important fragments of it).

It depends on circumstances whether it pays or not to try to reconstruct a theory as an elementary one, just like it depends on circumstances whether while looking for a solution to a mechanical problem it pays or not to apply certain statistical techniques. The "elementarization" of theories subjected to meta-theoretical investigations is a research method of a special importance. Its significance stems from the significance and strength of metamathematical results concerning elementary theories.

NOTE. Let $\mathscr{L}_\Theta$ be a language (defined in the usual way, i.e. as a fragment of the language of an extension of *ZF*) of an empirical theory Θ. Observe that the assumption that each Θ-structure determines a complete interpretation for $\mathscr{L}_\Theta$ rules out the possibility to refer $\mathscr{L}_\Theta$ to more than one Θ-structure at a time. Obviously one may be interested in comparing different Θ-structures or studying some relations among them.

The way in which this can be done is the following. Let $\mathscr{K}$ be the set of Θ-structures in which we are interested, and let $\varphi_1, \ldots, \varphi_n$ be variables selected in order to study the relations among Θ-structures in terms of them. Form the class of all structures of the form

$$(I, \varphi_1, \ldots, \varphi_n)$$

where I is an element of the power set $2^{\mathscr{K}}$.

Let us call structures of the sort described *power structures*, and the languages in which such structures can be studied *power languages*. The latter name will be applied especially in the case when the language to which it refers possesses, apart from the symbols of $\varphi_1, \ldots, \varphi_n$, the symbols of variables involved by the structures in $\mathscr{K}$. As a matter of fact, if the latter symbols do not appear in the language, it does not differ from the languages we discussed.

As an example of a theorem which can be stated only in a suitable power language let me quote the following (cf. Suppes, 1957, p. 302):

> "Every system of particle mechanics is equivalent to a subsystem of an isolated system of particle mechanics".

(Incidentally, this theorem being true when referred to the class of all PM-structures, need not be true when interpreted as a theorem about physical systems).

The relation of equivalence of two systems of particle mechanics (for the definition cf. Suppes, 1957) and that of being a subsystem are examples of parameters represented in the earlier discussion by symbols $\varphi_1, \ldots, \varphi_n$.

As another situation in which a power language becomes evidently necessary let me mention the invariance of mechanical laws under Galilean transformations of inertial frames of reference (*relativity principle*).

In spite of the obvious significance of the power language we shall keep our discussion restricted to the languages which do not fall under that category.

3.3. THE CONCEPT OF TRUTH

In what follows whenever a theory Θ will be dealt with we shall assume that both the language $\mathscr{L}_\Theta$ and the content of Θ are well-defined. It will be of minor significance whether Θ is defined by a set of explicitly stated axioms, or by the conditions imposed on Θ-structures (the implicit axioms for Θ), or perhaps still in some other way. Let us also assume that from now on we shall not pay attention to a rather subtle difference between the notion of a partial model for Θ and that of a Θ-structure: models for Θ will be freely referred to as Θ-structures and vice versa. Simply, we will not be interested in how Θ has been set up, in particular whether mathematical models for specific concepts are defined by the axioms of the theory or quite to the contrary, the definition of a model for the specific concept determines a certain axiomatization of a theory.

The language $\mathscr{L}_\Theta$ for Θ will be as usual assumed to be semi--interpreted, in particular some interpretations are defined to be standard interpretations for $\mathscr{L}_\Theta$, and moreover given such a structure I, the truth value of any sentence α of $\mathscr{L}_\Theta$ under that interpretation is defined.

A rigorous and at the same time satisfactory definition of the notion of validity need not be easily available, and still worse, it need not be available at all. The thing depends on the nature of both $\mathscr{L}_\Theta$ and its standard interpretations. Since we are not going to deal with any specific language $\mathscr{L}_\Theta$ and any specific definition of Θ-structure, we are not able to discuss the matter in detail. We can merely describe in an informal manner the main steps which must be taken in order to define the notion in question.

The reader who is familiar with Tarski's conception of truth should have no difficulty in following the subsequent considerations. In fact, most likely he will find them trivial. Still, for the sake of completeness, I decided not to resign from commenting on the notion of validity, the more so this concept will be of fundamental significance for the later discussion.

The steps to be taken towards defining or at least characterizing in an axiomatic way the notion of validity are the following.

STEP 1. Give a rigorous definition of $\mathscr{L}_\Theta$ within ZF, by defining a set $Form_\Theta$ such that:

(i) for each $x \in Form_\Theta$, ZF contains a constant symbol which denotes x,

(ii) there is an effectively defined one-to-one correspondence n between the set of sentences of $\mathscr{L}_{AR}$ and $Form_\Theta$.

Note that languages of the form $\mathscr{L}_\Theta$ are defined "outside" ZF, i.e. in an informal (or rather semi-formal) metalanguage in which ZF and all its extensions can be discussed. Obviously, this is that very (meta) language in which all our considerations are carried out, for instance in which the languages $\mathscr{L}_{AR}$ and $\mathscr{L}_{CT}$ were defined. (Note also that elements of $Form_\Theta$ are not assumed to be sentences by themselves, but in view of Clause (ii) they may be considered to be names for sentences of $\mathscr{L}_\Theta$).

STEP 2. Let I be an arbitrary standard interpretation of L_Θ. Denote by ZF_I the axiomatic extension of ZF which results from adding the symbol I as a new constant to the vocabulary of ZF and enlarging the list of axioms with some new ones which taken together state that I is a standard interpretation of $\mathscr{L}_\Theta$. Clearly, when we have to our disposal the symbol I, we may define in ZF_I the symbols denoting the constituents of I. In this way ZF_I becomes the language in which we may speak about I, although, paradoxically enough, in view of the assumption made the sentence "I is a standard interpretation for $\mathscr{L}_\Theta$" may represent all we know about I.

STEP 3. In the same metalanguage in which $\mathscr{L}_\Theta$ and ZF_I are defined, define a set of sentences $\mathscr{L}_I$ of ZF_I, relevant to I in the sense that their truth-values depend on the properties of I.

STEP 4. Define a one-to-one mapping $Trans_I$ of $\mathscr{L}_\Theta$ onto $\mathscr{L}_I$ so that from an intuitive point of view a sentence $\alpha \in \mathscr{L}_\Theta$ holds true in I if and only if $Trans_I(\alpha)$ holds true in I. The definition should be such that for each sentence α in $\mathscr{L}_\Theta$ there is an effective procedure by means of which the sentence $Trans_I(\alpha)$ can be constructed. The latter will be called the translation of α in ZF_I.

It is important to realize that the adequacy of the solutions to the problems one faces when trying to accomplish Step 3 and 4 can be judged only on the ground of certain intuitive assumptions, which need not be commonly shared. Moreover, it may happen that in view of the intuitive criteria we have adopted Step 3 and Step 4 cannot be carried into effect without violating the accepted adequacy criteria.

The main obstacle one may encounter is that the truth value

of some sentences of $\mathscr{L}_\Theta$ under the interpretation I may turn out to be dependent on something else than I. After all I provides an interpretation only for the specific symbols of $\mathscr{L}_\Theta$, while the language $\mathscr{L}_\Theta$ involves set-theoretical and perhaps also other mathematical constants. For illustration consider the axioms for AR. Unlike the theory CT, in the case of which the axioms restrict the range of variables to the set $\mathscr{L}$, the axioms for AR do not impose any confinement on the range of the variables appearing in them. This was the reason why AR1 – AR7 appeared as the clauses of the definiens of Definition 3.1.2 in a modified version. We replaced $\in$ and Z by $\in_U$ and Z_U, respectively. The structure

$$(Z_U, \in_U, \cdot \mid ', 0, S)$$

is called a set-theoretical superstructure for $(\cdot \mid ', 0, S)$. It turns out then that in order to decide whether the axioms for AR are true or not under a given partial interpretation $(\cdot \mid ', 0, S)$ we must first define a set-theoretical superstructure which will provide an interpretation for Z and $\in$.

The situation we have described is not characteristic of $\mathscr{L}_{AR}$ only. Note for instance that if the language $\mathscr{L}_\Theta$ includes sufficiently large part of the language of ZF it may happen that some of the sentences it comprises do not refer to any particular interpretation for $\mathscr{L}_\Theta$ but e.g. to all of them. For an example consider the sentence:

(*) *For each language $\mathscr{S}$ the set of all consequence operations on $\mathscr{S}$ forms a complete lattice.*

which may easily be formed in ZF. The expressions "language" and "consequence operation" are definable in ZF at least in such a general manner as it has actually been done (without using the expressions explicitly) in Definition 3.1.4. There are no general rules which prevent including sentences like (*) into $\mathscr{L}_{CT}$, but if one decides to do so, neither single CT-structures nor even single structures termed standard interpretation for $\mathscr{L}_{CT}$ (cf. Definition 3.2.5) provide sufficient interpretation for the symbols of $\mathscr{L}_{CT}$ defined in that way. In order to have the notion of validity applicable to all sentences of $\mathscr{L}_{CT}$, in particular to (*), it becomes necessary to define the notion of an interpretation for that language in a more involved manner.

STEP 5. Add a new predicate symbol Val_A to the vocabulary of ZF_A and complement the axioms for ZF_A with all the sentences which fall under the following schema:

(T) $Val_I(n(\alpha))$ if and only if α_I,

where $n(\alpha)$ $(n(\alpha) \in Form_\Theta)$ is the name of the sentence α of $\mathscr{L}_\Theta$, and $\alpha_I = Trans_I(\alpha)$.

Note that (T) is a variant of the well known Tarski's schema of a partial definition of truth (cf. A. Tarski, 1956, p. 155).

STEP 6. Check whether the axiomatic extension described above can be reduced to a definitional one, i.e. whether the symbol Val_I can be defined in ZF_I in such a way that the definition will allow to prove all sentences of the form (T). Accept such a definition if it is available.

In general the definition of the set Val_I need not exist. Certainly in that case we must content ourselves with the axiomatic characterization of Val_I, i.e. conclude our efforts at Step 5.

In order to illustrate the procedure described let us discuss briefly the way in which it can be applied in the case of *CT*. We shall not try to define the language $\mathscr{L}_{CT}$, let us assume however that it includes the following two sentences (which incidentally implies that $\mathscr{L}_{CT}$ considered now does not coincide with that we defined in the previous section).

(1) There is $\alpha \in \mathscr{L}$ such that $Cn(\alpha) = \mathscr{L}$,

(2) For all $\alpha, \beta \in \mathscr{L}$ there exists $\gamma \in \mathscr{L}$ such that $Cn(\alpha, \beta) = Cn(\gamma)$.

There are many ways in which the names for expressions of a language, in the present case $\mathscr{L}_{CT}$, can be formed. Perhaps the simplest one consists in defining the name for each compound expression as a finite sequence of the names for elementary expressions the former involves. We shall, however, stick to the convention which thus far has often proved to be useful, and depending on the context apply the same expression in the referential or selfreferential manner.

Select any standard interpretation for $\mathscr{L}_{CT}$. It may be assumed to be uniquely defined by a structure $(\mathscr{L}_0, Cn_0)$ such that:

(a) $\mathscr{L}_0$ is a non-empty set.

(b) $Cn_0\colon 2^{\mathscr{L}_0} \to 2^{\mathscr{L}_0}$.

Conditions (a) and (b) are axioms imposed on $(\mathscr{L}_0, Cn_0)$, they may exhaust everything we know about $(\mathscr{L}_0, Cn_0)$.

We have accomplished Step 2 and the next thing we have to do is to define sentences by means of which $(\mathscr{L}_0, Cn_0)$ should be characterized. The simplest possibility, which may happen to be the most reasonable one, is to select those sentences as direct counterparts of sentences of the language discussed, $\mathscr{L}_{CT}$ in the presented case. To begin with let us examine this possibility.

Suppose that the sentences which serve as translations of (1), (2) are:

(1′) There is $\alpha \in \mathscr{L}_0$ such that $Cn_0(\alpha) = \mathscr{L}_0$,

(2′) For all $\alpha, \beta \in \mathscr{L}_0$ there exists $\gamma \in \mathscr{L}_0$ such that $Cn_0(\alpha, \beta) =$
$= Cn_0(\gamma)$.

Consequently, the instances of schema (T) relevant to the example discussed are:

(T_1) $(1) \in Val_{(\mathscr{L}_0, Cn_0)}$ if and only if (1′),

(T_2) $(2) \in Val_{(\mathscr{L}_0, Cn_0)}$ if and only if (2′).

Instead of (1), (2) one should write the names for sentences (1), (2) (clearly symbols (1), (2) are also such names but selected in a contingent way) and instead of (1′) and (2′) the sentences (1′) and (2′), respectively. Obviously (1′) serves as $Trans_{(\mathscr{L}_0, Cn_0)}$ (1) and (2′) as $Trans_{(\mathscr{L}_0, Cn_0)}$ (2).

As long as conditions (a) and (b) exhaust our „knowledge" about $(\mathscr{L}_0, Cn_0)$, (T_1), (T_2) are of no much use. Suppose that for instance

(c) $\mathscr{L}_0 = \{0, 1, \ldots, 10\}$,

(d) for each $X \subseteq \mathscr{L}_0$, $Cn_0(X) = \{a \in \mathscr{L}_0 : a \leqslant \max(X)\}$.

One may easily check that under these assumptions the sentence

(1) becomes valid in $(\mathscr{L}_0, Cn_0)$, while (2) is not. One may also easily see that only some of Tarski's axioms for CT are valid under the interpretation defined, which in view of Assertion 3.3.7 means that it is not a CT-structure.

It is worth mentioning that translations of sentences of the language examined need not be their trivial counterparts as it has been in the discussed case. Suppose for instance that $(\mathscr{L}_0, Cn_0)$ is a CT-structure and let Ω_0 be the set of rules of inference which define Cn_0. Under this assumption we may define the following sentences to be translations of (1) and (2), respectively:

(1″) There exists $\alpha \in \mathscr{L}_0$ such that for each $\beta \in \mathscr{L}_0$ there is a proof of β from α by means of rules in Ω_0,

(2″) For each $\alpha, \beta \in \mathscr{L}_0$ there exists $\gamma \in \mathscr{L}_0$ such that (i) there is a proof of γ from $\{\alpha, \beta\}$ and (ii) there are proofs of both α and β from γ.

The definition of the concept of truth may then involve some nontrivial semantic analyses.

3.4. EMPIRICAL THEORIES

By an *empirical theory* we shall mean an applied mathematical theory. Although this definition seems to fit in very well with some sophisticated empirical theories (notably physical ones), its adequacy with respect to those which involve little mathematics or do not involve it at all is not so evident. Let me point out however that all sentences, including those of the conversational language, have some mathematical structure (they can be interpreted in a set-theoretical manner) and consequently, all theories can be viewed as mathematical, although perhaps completely trivial, ones.

Whichever way the notion of an applied theory might be understood we may take it for granted that in any case to apply a mathematical theory means to apply it to a class of empirical phenomena $\mathscr{K}$, in order to give an account of regularities they display. It is rather obvious that the same mathematical theory may happen to be applicable to many different areas: to phenomena $\mathscr{K}, \mathscr{K}', \mathscr{K}'', \ldots$ of entirely different nature. A set of equations which provides a model for some physical regularities may turn out to be useful in characterizing the behaviour of some economical systems, or some chemical reactions, or something else.

When Θ is assumed to be a theory of $\mathscr{K}$ we shall denote it as $\Theta_{\mathscr{K}}$. Theories $\Theta_{\mathscr{K}'}, \Theta_{\mathscr{K}''}, \Theta_{\mathscr{K}'''}, \ldots$ will be regarded to be different although their "mathematical structure" is the same. Given an (empirical) theory $\Theta_{\mathscr{K}}$, the set $\mathscr{K}$ will be referred to as the *scope* of $\Theta_{\mathscr{K}}$ and at the same time as a (*possible*) *range of applicability* of Θ. Elements of $\mathscr{K}$ will be called *domains* of $\Theta_{\mathscr{K}}$ and at the same time *possible domains* of Θ. Finally Θ will be referred to as a *mathematical formalism* of $\Theta_{\mathscr{K}}$. We have already mentioned that the same mathematical formalism may be characteristic of different empirical theories. Clearly, it is also possible that the same class of phenomena $\mathscr{K}$ is studied with the help of theories $\Theta'_{\mathscr{K}}, \Theta''_{\mathscr{K}}, \ldots$ based on different mathematical formalisms. (Heisenberg's and Schrodinger's formalizations of quantum mechanics may serve as an example).

In our further considerations we shall consistently try to view singular phenomena as abstract structures and thus scopes of theories as sets of similar structures of a certain fixed sort. In some cases, however, it seems advisable not to decide in advance what structures are to serve as models for phenomena examined by a theory in order to make it possible to admit the same singular phenomenon to be represented by systems of different sorts. Having such situations in mind let us now make a convention that sometimes the range of a theory will be understood to be a set of phenomena in the physical sense. Then the letter p provided with a subscript, if necessary, will serve as a variable representing a singular phenomenon in the physical sense. Systems serving as models for p will often be denoted by symbols provided with the subscript p, eg. $\mathfrak{A}_p$ or $\mathfrak{a}_p$.

It is quite clear that in order for the theory Θ to serve as a theory of phenomena in $\mathscr{K}$, the systems representing these phenomena have either to be partial interpretations for $\mathscr{L}_\Theta$ or to bear well-defined relation to the latter. Assume then that to each phenomenon $p \in \mathscr{K}$ a class $c(p)$ of partial interpretations for Θ is assigned. The rules which define c will be referred to as correspondence rules and the function itself will be called a correspondence function.

To give a better idea of what is to be meant by the correspondence function and how it is to be defined, let us consider a few examples. The definition of c becomes especially simple when the system $\mathfrak{A}_p$ representing p is a standard interpretation for $\mathscr{L}_\Theta$. In that case we shall put $c(p) = c(\mathfrak{A}_p) = \mathfrak{A}_p$. Another not much complicated case we can meet is when the elements of $\mathscr{K}$ are represented by reducts of partial interpretations for $\mathscr{L}_\Theta$. Clearly, if $\mathfrak{A}_p$ is such a structure, then we shall set $c(p) = c(\mathfrak{A}_p) = \{\mathfrak{B} \in Int(\mathscr{L}_\Theta): \mathfrak{A}_p$ is a reduct of $\mathfrak{B}\}$. Still another situation is faced when data on the phenomena in $\mathscr{K}$ concern quantities different from those involved by the theory. It is then necessary to define interrelations between empirical and theoretical terms in a suitable enlargement of the language $\mathscr{L}_\Theta$. The situation described is of course an analogon of that investigated in philosophical literature when interrelations between theoretical and observational languages are studied. The very term "correspondence rule" was coined just when interrelations between those two kinds of the language became subject to investigations (cf. e.g. P. W. Bridgman 1950, R. Carnap 1955, and 1966, E. Nagel 1961, H. Reichenbach, 1951).

Discussion to be found in the next chapter includes natural examples illustrating the last of the mentioned possibilities. Qualitative structures which will be discussed there may be treated as systems to which a theory formulated only with the use of quantitative terms relates. Thus the parameters involved in the characteristic of those structures vary from the variables involved in the theory.

At least two factors affect the choice of that and not some other kind of structures to be models for investigated phenomena. The theory with which the phenomena are examined forces a conceptualization of them. At the same time if structures serving as models for examined phenomena are to be something more than simply abstract invents, their properties must be determined by available empirical data. The structures ought to be defined "in close relation to experiment".

Considerable fragments of our further considerations, especially those undertaken in Chapter V will be devoted to explanations how the character of available empirical data may determine the choice of structures to be models for phenomena. Anticipating these considerations we shall try now to present in a sketchy way certain problems which may arise here.

Suppose that systems being elements of the scope $\mathscr{K}$ of a theory $\Theta_{\mathscr{K}}$ are reducts of partial interpretations for $\mathscr{L}_{\Theta}$ and assume that they are elementary systems of the form

$$\mathfrak{a}_p = (F_1^p, \ldots, F_n^p)$$

(such an assumption, at least as a starting hypothesis will be in force also in Chapter V). The functions F_i^p are thus functions of the form $F_i^p\colon T(\mathfrak{a}_p) \rightarrow Re$. If each sentence of the form $F_i^p(t) = x$ (where t and x are of course constant symbols) were empirically decidable, empirical procedures operated with would make it possible (at least theoretically speaking) to collect data determining a diagram of the system $\mathfrak{a}_p$. (By a *diagram of a structure* we mean the set of all atomic sentences valid in that structure which moreover is complete in the sense that it is built in a language containing symbols of all objects on which the parameters of the structure are defined; in case of $\mathfrak{a}_p$ the language must contain names for all instants from the interval $T(\mathfrak{a}_p)$ as well as names for all real numbers being values of the functions F_i^p. This very loose definition of a diagram is an adaptation of the notion known in logic, although usually defined only with respect to relational structures, cf. for instance A. Robinson 1963).

To define a diagram of the structure $\mathfrak{a}_p$ is equivalent to defining the structure in question. Thus in the case being discussed the relation between empirical evidence and the system which that evidence concerns is very simple. Unfortunately, in most cases the kind of relation we depicted turns out to be unrealistic simply because atomic sentences of the form $F_i^p(t) = x$ are not empirically decidable.

Very often the results of measurements of the quantities F_i^p can be represented only in the form of the sentences

(1) $F_i^p(t) \in \Delta$,

where Δ is an interval of real numbers (or, more generally, cf. Section 5.1, a Borel set). Thus the measurement is of imprecise character. Improvements of measurement techniques can enable better approximations of the "real" value of $F_i^p(t)$ (clearly, we need not assume that such a value exists in the physical sense; its existence may be treated as some useful convention). If however F_i^p is a continuous quantity, the values of F_i^p cannot be measured in a precise way. Consequently, no experiment can provide us with the diagram of $\mathfrak{a}_p$. What is experimentally available is at best an *approximate diagram of* $\mathfrak{a}_p$, i.e. a set D of sentences of the form (1) such that the following two conditions are satisfied:

(D1) If $F_i^p(t) = x$, then for some Δ such that $x \in \Delta$, the sentence '$F_i(t) \in \Delta$' is in D.

(D2) If '$F_i(t) \in \Delta$' is in D, then $F_i^p(t) \in \Delta$.

(Clearly, to say that $F_i^p(t) = x$ amounts to saying that '$F_i(t) = x$' is valid in $\mathfrak{a}_p$).

The procedures by means of which atomic hypothesis concerning the system $\mathfrak{a}_p$ are decided may happen to be not only imprecise but also dispersive, i.e. application of them not necessarily must lead to univocal results. Such a case being faced, the results obtained by means of the procedures with respect to the sentence (1) must be elaborated statistically, and the final result will take the form of an assertion stating that the probability that the value of the function F_i^p at t is in Δ is such and such. In greater detail the notion of a dispersive procedure will be discussed in Section 4.4. As the data collected in result of application of dispersive procedures are inconclusive, the problem of defining the conditions which the structure $\mathfrak{a}_p$ must satisfy to be treated as the one having properties consistent with the gathered empirical evidence becomes pretty complex. Regardless of that difficulty however, we must ask ourselves the question whether our decision to represent phenomena in $\mathscr{K}$ by elementary systems is the best one, i.e. whether some other kind of systems would not do better. Note that this question arises already when it turns out that a fully sharp measurement of the quantities constituting the characteristic of $\mathfrak{a}_p$ is impossible.

Let us observe here that any considerations on semantic relations between a theory and its scope are based on analyses of empirical data which are certain linguistic (syntactical) objects. This may suggest the standpoint that only syntactical analyses may be considered to be well-defined, whereas semantic ones inevitably bear the tinge of speculation. This viewpoint is intimate for all those who are operationalistically-minded.

I am afraid that any attempt to prove that semantic account of science enables deeper approach to methodological problems than the syntactic one would be a failure. The problem lies rather in the choice of research methods, and it is final results and not verbal arguments which decide whether decisions made were right or not. Having decided on the semantic approach, we certainly must not feel relieved from the necessity to confront it all the time with the syntactic one.

The procedure which we are going to accept in order to confront the two approaches, is simple. We assume that to each domain p of a theory $\Theta_{\mathscr{K}}$ there corresponds a class of sentences H_p which are empirically decidable, although not necessarily in a conclusive way (empirical procedures applied need not be dispersive free). Hypotheses in H_p need not on the whole be sentences of the language $\mathscr{L}_{\Theta}$, they may belong to an enlargement of this language. If the theory Θ is to be applicable to $\mathscr{K}$, relations between Θ and sentences in H_p must be well-defined. Thus for instance if terms occurring in empirical hypotheses are not the terms involved by the theory, we must have at our disposal certain rules which define how theorems expressed in terms of one of the two kinds yield theorems stated in terms of the other. Clearly those are simply correspondence rules, although treated as syntactic ones. I hope the schema of the procedure outlined above is clear, though certain technical problems which may arise in its realization need not be simple. Those are however troubles we shall try to resolve later (cf. Chapter IV).

3.5. TWO EXAMPLES OF EMPIRICAL THEORIES

The objective of this section is to illustrate the discussion carried out in the previous section by suitably selected examples. For this purpose we shall exploit two of the theories characterized earlier (cf. Section 3.2), namely the theory of linear hierarchy pattern *LH* and particle mechanics *PM*.

What is lacking in the formal exposition of *LH* and therefore must be complemented at least by suitable comments, is a characteristic of the intended range of applications of the theory, let us denote it as $\mathcal{K}$. To put it in other words, we have defined a mathematical formalism *LH*, but we want to have a definition (a least a rough one) of an empirical theory $LH_{\mathcal{K}}$.

The linear hierarchy pattern is observed in various animal groups, in particular in small groups of hens which lived together for a relatively long time. Instead of trying to set up $\mathcal{K}$ in a fully general way let us restrict our discussion just to groups of hens only, thus to a subset Kh of $\mathcal{K}$.

One might try to be more specific about what is to be meant by "a small group" and a "relatively long time", but it is quite evident that no rigorous definition of these locutions is easily available. Anyway, what matters is that if a group is too large, some of the individuals in it may contact one another too rarely for a domination pattern to become fixed. Similarly, too short period of being together is another obstacle to a complete settlement of a hierarchy in the group. Clearly the period necessary for settling the hierarchy depends on the size of the group, but it depends also on some other parameters, for instance the living conditions in which the hens are kept.

The difficulties we mention affect the possibility of defining the set of all intended interpretations for the symbol G in a clear-cut way, but we may adopt the following policy. Let us form a sequence

(1) $$\Xi_1, \Xi_2, \ldots, \Xi_i, \ldots$$

of sets of groups of hens, such that

(i) for all i, $\Xi_i \subseteq \Xi_{i+1}$, and

(ii) Given any Ξ_i in the sequence, and given any G in Ξ the suitably trained experimenter will recognize G as a group of hens to which LH is applicable.

(iii) The set $\Xi = \bigcup \{\Xi_i : i \in \mathcal{N}\}$ comprises all groups of hens which determine domains of LH_{Kh}.

Clearly Ξ should be treated as a limit concept which cannot be defined in any satisfactory manner but merely approximated by subsequent Ξ_i. Still, if one is not afraid of taking such a step he may assume that Ξ is well defined, and introduce and use it freely in the discussion carried out. As a matter of fact we did that when accepting Clause (iii). It is worth remarking, however, that Clause (iii) may be dropped out. The one who is constructively--minded, and is not eager to accept any assumption on existence of a set which cannot be defined in an effective manner, may restrict his considerations to finite sequences of the form (1), without pressupposing what the greatest n for which Ξ_n still is an element of the sequence might be. In that way instead of dealing with one intended scope Kh he starts dealing with sucessive elements of the sequence

$$Kh_1, Kh_2, \ldots, Kh_i, \ldots,$$

each Kh_i being a set of abstract structures determined by the corresponding set of groups of hens Ξ_i.

Let us also comment briefly on Condition (i). One should perhaps avoid introducing pragmatical elements (i.e. a reference to the users of the language who in (ii) are represented by the experimenter) to semantic considerations of the sort carried out now. But there is nothing wrong with assuming that the final decision on whether LH is applicable to a particular group G or not depends on the expert's opinion. One obviously may try to analyze the way in which experts take their decision, and for instance list the criteria they apply. It may turn out however that a complete account of the procedures which serve to discriminate a genuine domains of the theory from alleged ones cannot be provided in any sufficiently common language since such an account may involve references to some attributes or situations recognizable only by the skilled experimenter and having no names for them being in common use. The experimenter may invent (or even may possess) suitable names, but they would be names of his private language.

We have devoted quite a lot of space to the problem of interpreting the symbol G, let us in turn eaxmine the same problem relative to U.

The most natural interpretation of U is the one under which U includes all whatever might be targets of competition among the individuals in the group. But such a decision leaves a wide margin of indefiniteness as to what U amounts to and although this does not undermine the possibility of applying LH, it still limits, or even rules out, the possibility of examining some of the methodological properties of LH. In particular, soundness of LH with respect to a particular application domain $\mathfrak{G}$ depends on how U is defined.

The best way to cope with the problem seems to be the one analogous to that we described when discussing interpretation of G. Namely we may assume that

$$U = \bigcup \{U_i : i \in I\},$$

where

$$U_1, U_2, \ldots, U_i, \ldots$$

is a sequence of finite sets of targets such that $U_i \subseteq U_{i+1}$, for all i. Again one may decide to deal with U and treat U_i's as finite approximations of U, or assign to each U_i a separate domain $\mathfrak{G}_i$ of LH_{Kh}. Obviously in practice we may examine the hierarchy pattern only with respect to a fixed and not too numerous set of targets, which corresponds to the latter decision, but at the same time we are not likely to assume that the results of our inquiry are relevant only to that particular U_i we have dealt with.

There are two more symbols, D and P, which need to be interpreted. We shall assume that D is a theoretical symbol, i.e. has no "direct" empirical interpretation. The domains of LH_{Kh} will be then viewed as structures of the form

$$\mathfrak{G} = (G, P, U),$$

where for each $X \subseteq G$ and each $u \in U$, $P_{X,u}$ is a probability measure and thus P may be conceived as a probability measure depending on two parameters: $X \subseteq G$ and $u \in U$. The value of $P_{X,u}(a, k)$ should be estimated in terms of relative frequencies in data concerning subsequent trials.

The theory described here is very simple, which from our point of view it its adventage. It is to serve only to illustrate the considerations which are carried out. I find certain properties of the theory LH_{Kh}, some of them already discussed, specially worth exposing. They are:

(i) The range of possible applicability of LH_{Kh} consists of a great number of different systems which would be difficult to combine into a single domain of LH. This obvious remark is not so redundant as it may seem to be since it is very often taken for granted that each empirical theory refers to a "fragment" of a real world, an thus it has exactly one intended interpretation. I do not claim that in no case whatsoever elements of Kh could be combined into a certain single structure. But it would not be an easy thing to do if not for any other reasons then because Kh is not, a strictly defined class. Also the expediency of combining elements of Kh into a singular system is doubtful. In all practical applications the theory LH is referred to a single group of animals living in a certain specific period of time on a certain specific territory.

(ii) The quantity P (more precisely its specifications $P_{X,u}$) is a probability measure. None the less it may be treated analogously to quantities of nonstatistical character. The theories to which our considerations refer may involve statistical notions. However the notions must be defined with respect to particular domains of the theory and not to its whole range.

(iii) The theory LH_{Kh} includes a "theoretical" parameter – the relation D. We shall not apply the term "observational" in our considerations. But to some extend in our discussion empirical terms will play the role of observational ones. What discerns the partition of terms into theoretical and empirical from that into theoretical and observational is that we assume that empirical terms may be heavily "theoretically laden", i.e. they need not represent the direct pure experience, what observational terms are notoriously expected to do.

The partition of terms into theoretical and empirical is based on intuitions closer to those which in physics determine the partition of quantities into measurable and non-measurable rather than to those which have been fundamental for the theoretical-observational dychotomy. Measurable quantities need not be observational.

In fact a careful examination of contexts in which the notion of a theoretical term is applied reveals that it appears at least in two meanings. Sometimes "theoretical" means non-observational, but more often the term "theoretical" acquires a much narrower interpretation. Roughly speaking a theoretical term in the latter sense is a term which is out of place in reports on results of experimental inquiry; the scientist who wants to present "facts" (empirical data gathered by standarized commonly accepted methods) should refrain from using it.

(iv) It is customary to treat probability as a certain formal construction rather than a genuine physical quantity, such as say length or electric charge, and consequently to take it for granted that the symbol P appearing in LH possesses only an "operational" meaning, i.e. the denotation of P is defined to the degree to which the methods of measurement of P are defined. This strikingly operational nature of P introduces a desired element of balance to our prevailingly semantic approach and calls attention to the fact that any deeper analysis of relations between empirical theories and their domains must inevitably lead to the problem of definitional role of testing procedures.

Let us exploit the example of an empirical theory provided by LH_{Kh} and define the conditions under which LH_{Kh} may be considered to be sound. The definition we are going to state should provide a useful illustration for the later, more involved, and more systematic discussion of the notion of soundness.

Given a system $\mathfrak{G} \in Kh$, denote by $c(\mathfrak{G})$ the class of all possible expansions of G being standard interpretations of the language

$\mathscr{L}'_{Kh}$. The function c is of course the correspondence function discussed in the preceding section. The definition below is self-explanatory.

DEFINITION 3.5.1. *The theory* LH_{Kh} *is* semantically sound in the system $\mathfrak{G} \in Kh$ *if and only if there exists a standard interpretation* $I \in c(\mathfrak{G})$ *such that* $LH_{Kh} \subseteq Val(I)$.

Definition 3.5.1 involves tacitly the assumption that each group of hens in Kh can be viewed as an abstract well defined system. Observe however that the system $\mathfrak{G}$ to which the definition refers is not defined univocally since an univocal definition of P being one of the variables of the system is not possible. The values of P can be measured and thus defined with restricted accuracy only.

Consequently from the operational standpoint at least some of the sentences of the form $P_{X,u}(a, k) = x$ may have no empirical meaning at all, as they do not submit to empirical verification. In certain special situations one can perhaps establish that $P_{X,u}(a, k) = 0$ (for some particular and specified X, u and k!) or that $P_{X,u}(a, k) = 1$, but no experiment can suffice to establish that, say, $P_{X,u}(a, k) = 1/2$. We can merely estimate the value of $P_{X,u}(a, k)$, i.e. the results of experiments will take the form:

(2) $$P_{X,u}(a, k) = x \pm \varepsilon,$$

in general with $\varepsilon > 0$.

Let us briefly comment on how hypotheses of the form (2) are tested.

A typical statistical experiment consists of a series of trials. The experimenter's task is to establish the relative frequency of the outcomes of the trials which belong to the event in which he is interested, and next estimate the probability of the event. The last step however by no means amounts to trivial calculations. In order to estimate the probability of an event with the help of relative frequencies, one must make use of some "a priori" assumptions. The reason why such assumptions are necessary is quite obvious. Since we are not able to produce infinite series of trials we must evaluate the probability by analyzing the data concerning the finite series we are able to produce. But a finite experiment can never suffice to decide all questions relevant to the tested hypothesis. In particular no finite experiment allows to decide whether outcomes of the trials are mutually independent, simply because the set of all patterns of possible correlations among outcomes of particular trials is infinite.

The best way to cope with the difficulty is to accept such a model of the relation between probability measures and relative frequencies which conforms to both the theory concerning the events examined (in our example *LH*) and common sense, and which at the same time is as simple as possible, and just start working with that model until something makes us reconsider our decision. At least that is the way in which the difficulty would be solved in practice. By a model of relations between probability measures and relative frequencies I simply mean a set of equations which allow to transform empirical evidence stated in terms of relative frequencies onto sentences of the form (2).

Denote as H_G the set of all hypotheses of the form (2) referring to G which are empirically testable, and denote as τ_G the function which assigns to each α in H_G one of the two values 1 and 0 depending on whether the experiment confirms or falsifies the hypothesis. Put also

$$E_G = \{\alpha\colon \tau_G(\alpha) = 1\} \cup \{\neg\, \alpha\colon \tau_G(\alpha) = 0\}.$$

Finally, assume that τ stands for the set of all testing procedures applied when empirical hypotheses relevant to LH_{Kh} are verified. The function τ_G can be then viewed as being defined by τ.

DEFINITION 3.5.2. *The theory* LH_{Kh} *is* deductively (semantically) sound in $G \in Kh$ *if and only if the union* $LH \cup E_G$ *is deductively (semantically) consistent.*

In contradistinction to Definition 3.5.1 the present definition does not presuppose existence of any abstract system $\mathfrak{G}$ which represents the group G. We have stressed this by writing deliberately "$G \in Kh$" instead "$\mathfrak{G} \in Kh$" which certainly implies that the symbol Kh is applied in two different ways: it denotes either the set of abstract systems or the set of their epirical counterparts (groups of hens). Incidentally, the ambiguity of that sort will be characteristic of most of our considerations in which the notion of the scope of a theory is involved.

Now observe that as soon as we accept Definition 3.5.2, we begin to treat the theory dealt with as the triple $(LH, \tau, \mathscr{K})$. The testing procedures τ gain rights equal to those of theorems of the theory, in particular soundness of the theory ceases being the matter of interrelations between LH and elements of Kh only and becomes dependent on all the three constituents determining the theory.

Note that it may happen (none of the assumptions made excludes this) that already the set E_G itself is contradictory, not to mention $LH \cup E_G$. Clearly, in such a case there is no doubt that it is the procedures τ which yield inconsistent results and thus they must be changed before one starts changing anything else. But even if testing procedures do not yield inconsistency, the fact of existence of divergencies between the theory and empirical evidence does not provide a conclusive argument in favor of changing the theory rather than the testing procedures applied.

Perhaps the discussion of the notion of soundness of LH_{Kh} given above should be complemented by some comments on how Definitions 3.5.1 and 3.5.2 are interrelated. But this question involves certain subtleties concerning the notion of the correspondence function c into which I do not want to go just now. The thing will be discussed in a general manner (i.e. without reference to any particular theory) in Chapter V (cf. Assertion 5.4.1 and remarks relevant to it). After all, all that has been said about LH_{Kh} has been meant to serve only as an illustration for more general considerations to be given later (for instance Definition 3.5.2 provides an exemplification of Definition 5.2.2), and going into too many details at the present, preparatory stage of the discussion would be, I am afraid, inappropriate.

Since LH_{Kh} is not only a very simple but also artificial theory, let us discuss briefly a more sophisticated example. Let us examine how the range of applicability of particle mechanics can be defined and how particle mechanics is related to physical systems being its possible domains.

It is very unlikely that someone might be able to define the set of all physical systems to which *PM* can be applied. Rather, analogously to what was done in the case of *LH*, we may try to define certain selected areas of applications in hope that combined together they will comprise all systems which may be reckoned as domains of the theory. The analogy is not complete, however. The conceptual apparatus of the theory of linear hierarchy is adjusted to applications in many different areas. Without any enlargement of the list of specific concepts of the theory one may apply *LH* to studies of the behaviour of hens and, say, the behaviour of bisons. The changes which such a shift must imply are of semantic not syntactic nature: one must reinterpret the symbols *G* and *U*.

In the case of particle mechanics the way of complementing the list of specific terms of that theory must vary from application to application. The key problem we face when trying to use *PM* as a theory of a mechanical phenomenon of a special kind (say a free movement of a singular particle) is to define the Lagrangian in the way appropriate for that particular sort of application. In order to do that, especially in order to do that in such a way that the operational interpretation of the theory becomes transparent, we must, as a rule, introduce some new symbols and define Lagrange's function *L* in terms of them. *PM* should be viewed then as a sort of "super-theory" which only after suitable complementation becomes an applied theory provided with a homogeneous class of domains (cf. 1.6).

Let me illustrate these remarks by a short discussion of so-called isolated and constraint free systems of particles, i.e. systems whose behaviour (movements of the bodies the system involves) is not affected by anything from the outside of the system. In that special case the Lagrangian function *L* takes the form

$$L = \sum_{a \in A} \frac{m_a \mathbf{v}_a^2}{2} - U(\mathbf{s}_a, \mathbf{s}_b, \ldots), \tag{3}$$

where A is the set of all particles of the system examined, m_a denotes the mass of the particle a, $\mathbf{v}_a^2 = \dot{x}_a^2 + \dot{y}_a^2 + \dot{z}_a^2$ – where x_a, y_a, z_a

are Cartesian coordinates of the particle a, $\mathbf{s}_a = (x_a, y_a, z_a)$ is the so-called *radius vector* (*position*) of the particle a. Clearly the Cartesian coordinates $x_a, y_a, z_a, x_b, y_b, z_b, \ldots$ play the role of the generalized coordinates of the system $\mathfrak{A}$.

To have L defined one must specify the function U. The definition of that function depends on the nature of interactions of the particles in the system. If, for instance, they are electrically charged particles, the definition of U will involve the variable "electrical charge" and perhaps also some additional ones. The original language of *PM* must then be enlarged although that does not mean that the new variables will explicitly intervene in the definition of U (and thus L) in each case. For example, if electrical charges of the particles of a system $\mathfrak{A}$ are constant, the definition of U for $\mathfrak{A}$ may involve only the values which the parameter "electric charge" takes for particular particles without any explicit reference to that parameter. Incidentally, note that already Equation (3) involves the symbol m which did not appear in Axioms M1-M6 (cf. Definition 3.2.3).

We have no reason to go deeper into technical details characteristic of particle mechanics. After all, we are interested in methodological not physical matters. Let us assume that L is definable in $\mathfrak{A}$ in terms of the following variables: Cartesian coordinates x_a, y_a, z_a of a particle a, the first derivatives $\dot{x}_a, \dot{y}_a, \dot{z}_a$ of those coordinates, and mass m_a. For a system of N-particles we have then $7 \times N$ variables whose values need not be definable in terms of the remaining ones. Consequently, the state of the system $\mathfrak{A}$ at a given time t may be defined as a $7 \times N$-tuple: $\mathfrak{A}(t) = (m_a(t), m_b(t), \ldots, x_a(t), \ldots, \dot{x}_a(t), \ldots)$.

Let us denote the specification of *PM* we have discussed above as *PM**. One certainly may want to define *PM** in a formal manner, i.e. to define the notion of a *PM**-structure.

DEFINITION 3.5.3. *A structure* $(A, K, T, m, x, y, z, \dot{x}, \dot{y}, \dot{z}, U)$ *is a* PM*-structure *if and only if the following axioms are satisfied*:

Axiom 1. *A is a finite non-empty set of points of the 3-dimensional Euclidean space* $\mathscr{E}^3$, *called* particles. *The coordinates of particles depend on time.*

Axiom 2. *K is a Cartesian frame of reference for* $\mathscr{E}^3$.

Axiom 3. *T is an interval of real numbers (time interval).*

Axiom 4. x, y, z *are coordinates of the points in* A *defined relative to* K.

Axiom 5. $\dot{x}, \dot{y}, \dot{z}$ *are the first derivatives of* x, y, z, *respectively, with respect to the time variable* t.

Axiom 6. U *is a function of the form* $U(\mathbf{s}_a, \mathbf{s}_b, \ldots)$, *where* $\mathbf{s}_a, \mathbf{s}_b, \ldots$ *are the radius vectors of particles* $a, b, \ldots$ *in* A, *i.e. the values of* U *are defined by the values of the vectors* $\mathbf{s}_a, \mathbf{s}_b, \ldots$ *(or equivalently: by the values of the coordinates* $x_a, y_a, z_a, x_b, \ldots$*).*

Axiom 7. *For each* t *in* T, *the radius vectors* $\mathbf{s}_a(t)$ *are uniquely determined by the equation:*

$$m_a \frac{d\mathbf{v}_a}{dt} = -\frac{\partial U}{\partial \mathbf{s}_a},$$

where $\mathbf{v}_a = (\dot{x}_a, \dot{y}_a, \dot{z}_a)$, *provided that for some* $t_0 \in T$ *the values of vectors* $\mathbf{s}_b(t_0), \mathbf{v}_b(t_0)$ *are given for all* $b \in A$.

It is not immediately seen how the notion of a *PM**-structure is related to that of a *PM*-structure; in order to bring to light the correspondence between the two sorts of structures some mathematical transformations are necessary. What matters is that each *PM**-structure determines in a unique way a *PM*-structure and hence indeed the theory of *PM**-structures deserves to be called a *specification of PM*.

Let us call the reader's attention to the fact that $-\dfrac{\partial U}{\partial \mathbf{s}_a}$ is the force exerted on the particle a by the remaining particles of the system, and hence the equations Axiom 6 involves are the well known Newtonian laws of motion. Assume that the only source of forces to which the particles are subjected is the gravitational attraction among them. In that particular case we might complement the axioms for *PM** with the following one:

Axiom 8. $-\dfrac{\partial U}{\partial \mathbf{s}_a} = \sum\limits_{b \in A - \{a\}} \mathbf{g} \dfrac{m_a m_b}{r_{ab}^2}$, where **g** is a constant parameter (the *gravitational constant*) and $\mathbf{r}_{ab}^2 = |\mathbf{s}_a - \mathbf{s}_b|^2$.

Clearly, by enlarging the list of axioms for PM^* we arrive at a strengthening (and thus a specification) of PM^*. Let us denote it as PM^*_e. Since there are many ways of defining U (the definition depends on the nature of interactions between the particles), there are many theories of the form $PM^*_{e_1}, PM^*_{e_2}, \ldots$ being specifications of PM^*. We have then a hierarchy of theories which graphically can be presented as follows:

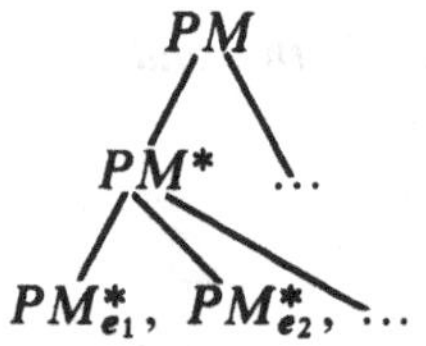

Only the theories which are located at the bottom part of the diagram are subject to direct applications.

The problems which arise when ones tries to define the range of applications for a theory of the form $PM^*_{e_i}$ are not substantially different from those we discussed in the case of LH. The question of whether LH is applicable to such and such group of animals living in such and such conditions is now replaced by that of whether $PM^*_{e_i}$ is applicable to such and such set of physical bodies which are in such and such conditions? In both the cases the experts have some criteria to decide the matter and in both the cases a satisfactory wording of those criteria is, I am afraid, impossible.

A typical domain of PM^*_e is the solar system. One certainly must decide in terms of what parameters the movements of planetery bodies are to be examined. Clearly, one of the possibilities is to take the parameters involved in the definition of the PM^*_e-system, but considerations of a sort similar to those which made us drop out the parameter D when defining empirical domains of LH may also this time lead to a similar decision. Obvious candidates for removal are U and the derivatives $\dot{x}, \dot{y}, \dot{z}$.

Assuming that K and T are fixed, we may then present domains of PM_e^* in the form

$$\mathfrak{A} = (A, m, x, y, z).$$

Let me emphasize once more that the decision to include some parameters into the conceptualization of empirical systems must reflect actual experimental possibilities rather than any ontological assumptions. The arguments which are to show that, say, just m, x, y, z are intrinsically empirical while $\dot{x}, \dot{y}, \dot{z}, U$ are intrinsically theoretical do not deserve to be trusted. We may easily figure out experimental devices which allow us to measure velocities and only then calculate positions rather than to apply the converse procedure, though certainly the latter is a standard one. The argument that force is a defined parameter and thus cannot be subjected to any direct measurement has often been met with scientists. On the other hand it is often claimed that it is mass not force which is defined by the third Newton law in terms of the remaining parameters the law involves. The point is however that all these parameters are involved in many physical laws, and some of them may provide a sufficient theoretical basis for measuring force or mass. The third Newton law is by no means the only one which may serve this purpose.

At that very moment when the notion of a possible domain of PM_e is defined, one may imitate Definition 3.5.1 in order to define the conditions under which PM_e should be considered to be sound in a given system $\mathfrak{A}$ to which the theory PM_e may be applied. One may also adopt the same approach which led us to Definition 3.5.2 and define the conditions of soundness of PM_e in a particular application in terms of consistency of the content of the theory and empirical evidence available. I shall not try to be more specific on the matter, since the problems to which the definition of soundness of PM_e may give rise do not differ essentially from those considered in the case of LH_{Kh}.

Let us conclude then this section with a few remarks concerning the two meanings in which one may claim that a theory is applicable to an empirical system.

By saying that Θ is applicable to $\mathfrak{A}$ one may inform that

Θ is sound with respect to $\mathfrak{A}$. Applicability understood in this way will be occasionally referred to as *actual applicability*. On the other hand, when speaking about applicability of Θ to $\mathfrak{A}$ one may have in mind that Θ is thought to be a theory of $\mathfrak{A}$ or, to put it in other words, that Θ is expected to provide a correct description of the structure or behaviour (or both) of $\mathfrak{A}$. The applicability understood along this line will be referred to as *intended applicability*.

It is obvious that whenever an applied theory $\Theta_{\mathcal{K}}$ is considered, all its domains are expected to be systems to which Θ is actually applicable. Thus if we found out that $\mathfrak{A}$ meets all criteria applied in order to discriminate the domains of $\Theta_{\mathcal{K}}$ from the systems which do not fall under this category, but on the other hand it does not behave according to some of the laws of the theory, we would have either to modify Θ or to reconsider the adequacy of the criteria. If the theory $\Theta_{\mathcal{K}}$ is well established, i.e. it has proved to function well (in a satisfactory way) in numerous applications, then the fact that $\Theta_{\mathcal{K}}$ is not sound relative to $\mathfrak{A}$ may make us reconsider the question whether $\mathfrak{A}$ deserves to be treated as a domain of the theory Θ rather than discuss the possibility of modifying $\mathfrak{A}$.

Laws of a theory often serve directly as criteria defining a scope of the theory. After random observations of the system $\mathfrak{A}$ which result in finding out that in the observed time-intervals the system behaves in accordance with laws of Θ one may feel justified to risk the conclusion that $\mathfrak{A}$ is in the scope $\mathcal{K}$ of $\Theta_{\mathcal{K}}$. A good basis for a medical diagnosis may be given by observation of an illness in its initial stage. Our reasoning is then as follows: according to our knowledge (a certain theory) the stages $c_1, c_2, \ldots, c_k$ occur successively in the course of the illness I. With a patient p the illness in the time of observation is consistent with the theoretical characteristic of c_1 and therefore we deal with the illness I.

The law stating that the course of I is of the form $c_1, c_2, \ldots, c_k$ (in given conditions, for example when the illness is not being cured) constitutes a basis for a diagnosis of the instances of the illness I, that is, domains to which the law applies. Moreover, if the empirical material collected is large enough we are able to estimate the degree of reliability of the diagnostic method described.

Let us turn back to particle mechanics. Laws of *PM** apply to inert systems. At the same time the best way for verifying whether a system is inert is to carry out a series of experiments in hope that they will show, though obviously in a non-conclusive way, that the system is subject to the laws of mechanics. In some textbooks on physics an inert system is defined to be a system which is infinitely distant from other material systems. This might serve as a criterion of the inertiality independent of *PM** but, I am afraid, the definition is not correct. This is not because it contains the word "infinitely", since in practical application it means that because of the distance the material influence of other systems is negligibly weak. But even the most remote and affected by the weakest external influence system need not be inert. Thus for example a rocket (treated as a particle) which moves in the space vacuum with working engines is not such a system when it speeds up or changes its trajectory.

A dual character of laws of empirical theories consisting in the fact that at the same time they define and describe a scope of a theory is one of the substantial features of empirical theories and its effect can be seen in many different situations. Notice, for example, that by its very nature an empirical theory has to be based on empirical data which justify and at the same time verify the theory. However, when following carefully the process of data collecting, one usually notices that laws of the theory are the laws on which measurements of the quantities specific for the theory are based. Thus we are faced again with some sort of a vicious circle being on analogon, if not a consequence, of the above mentioned twofold role of laws.

3.6. MODELS AND THEORIES OF EMPIRICAL PHENOMENA

There are three different, although in a way related meanings in which the term "model" has been applied thus far. In Section 1.6 we informally referred to abstract systems as models of empirical phenomena. In Section 1.9 a model was defined to be an interpretation for a language under which some sentences of that language become true. In that sense, i.e. the semantic one, "model" is a synonym of "realization". Clearly, the same system may happen to be a model for an instance of an empirical phenomenon and at the same time a model for a set of laws which describe the behaviour of that phenomenon. In Chapter II the term "model" appears in the context "a mathematical model of a regularity". A mathematical model in that sense has been defined to be a set of mathematical equations which describe some regularity in the behaviour of a phenomenon.

The list of meanings in which "model" appears in methodology of science and in empirical sciences is still longer. For example when two physical systems display certain correspondence in their behaviour, say they are just isomorphic, any one of them can be treated as a (physical) model for the other. If such a decision happens to be taken, the observations and experiments concerning the model are counted to be a source of information about the "prototype".

It might be desirable to reduce all, but perhaps informal, applications of the word "model" to one selected. On the other hand the ambiguity of the term "model" regardless of how untolerable it may seem to be, need not lead to misunderstanding, provided that some simple precaution measures are observed. Quite often the intended meaning of the term can be easily grasped from the context, if not – a brief explanation should suffice to remove doubts. The notorious ambiguity of the term "model" is well known but we learned somewhow to live with it. It is worth noting that while the semantic concept of a model is in common use among logicians and mathematicians working in the field of foundations of mathematics, empirical scientists almost unanimously opt for the "mathematical" meanings of "model" in its *syntactic* (a set of equations) or *semantic* (a mathematical entity which is to represent an intuitive concept) sense.

We are not going to discuss different meanings of notoriously ambiguous term "model" a penetrating discussion of such sort is to be found in P. Suppes (1961). Our objective will be to compare the syntactic concept of a mathematical model with that of a theory in the deductive sense. The áffinity of these two concepts is striking.

The question may be raised whether the notion of a mathematical model in the sense in which we are interested is not redundant. The equations which are to describe the behaviour of certain empirical systems are some empirical hypotheses. Thus setting up a mathematical model for a phenomenon amounts to setting up a theory of that phenomenon. This argument is especially compelling when a model provides a rigorous account of what has earlier been examined in a rather imprecise language.P. Suppes (1970a, pp. 2–12) describes this situation as follows:

> "A certain theory is stated in broad and general terms. Some quantitative experiments to test this theory are performed. Because of the success of these experiments, scientists interested in more quantitative and exact theories then turn to what is called the 'construction of a model' for the original theory. In the language of logicians, it would be more appropriate to say that rather than constructing a model they are interested in constructing a quantitative theory to match the intuitive ideas of the original theory".

There is no doubt that there is strong affinity between the notion of a mathematical model and that of a theory, as well as there is no doubt that in certain contexts in which the term "model" is used it would be more appropriate to speak about a theory. On the other hand, however, there is an evident need for a term which both when a theory is being constructed and when it is finally set up can be applied to denote those parts of the theory that provide a relatively complete account of particular regularities of phenomena in the scope of the theory.

From the formal standpoint one can treat any set of hypotheses or even any single hypothesis as forming a "theory". There is nothing surprising however in the fact that scienticists do not like to use the term "theory" in that way. An intuitive connotation of the term "theory" allows to apply it only to sets of hypotheses which are relatively well established (i.e. well confirmed both by experiments and applications) and give account of a rather large class of phenomena. Note also that a model can be inconsistent with both empirical facts and accepted theories, for instance it may be conceived at the very beginning as a construction which is merely to suggest a new way of looking at certain problems (heuristic models).

But there is another perhaps more serious reason why in some contexts one may prefer to use the term "model" rather than "theory". A set of laws which provides a relatively complete account of a certain regularity need not be autonomous in the sense that it may involve concepts whose theoretical meaning (i.e. the relations which they bear among themselves and to the other terms of the theory, if any) can be analyzed only in a broader context than that provided by the laws. For instance it would be rather artificial to divide particle mechanics into particle mechanics of one particle, of two particles, three and so on, although particle mechanics involves equations which serve as models for one particle system, two-particle system, etc.

One must admit, however that there is no clear-cut borderline between theories and models which conform to the intuitive meaning of the two words. Thus for instance although physicists often prefer to speak about Bohr's model of the atomic structure, the designation "Bohr theory of the atomic structure" is perfectly acceptable and, as a matter of fact, often used.

There is still an argument of a quite different nature to the effect that the syntactic notion of a mathematical model is redundant. If two different systems of equations are equivalent they form two different models of the regularity they describe. Observe now that each modification of equations even if it is entirely inessential (say writing in a new order the components of an operation invariant under such an alteration) results in changing the equations and thus in producing a new mathematical model. It seems more reasonable then, when speaking about mathematical models, to speak about mathematical entities the equations define which serve as abstract representations of the phenomena examined rather than about the equations themselves.

Let us ponder further on how the concept of a model meant to be an abstract representation of a phenomenon, the concept of a mathematical model and that of a theory are interrelated among themselves. The following quotation is an excerpt from W. R. Ashby's *Introduction to Cybernetics*, 1956.

"Suppose we have before us a particular real dynamic system – a simple pendulum, 40 cm long. We provide a suitable recorder, draw the pendulum through 30° to one side, let it go, and record its position every quarter-second. We find the successive deviations to be 30° (initially), and −24° (on the other side). So our first estimate of the transformation, under the given conditions, is

$$\downarrow \begin{matrix} 30^\circ & 10^\circ \\ 10^\circ & -24^\circ \end{matrix}$$

Next, as good scientists, we check that transition from 10°: we draw the pendulum aside to 10°, let it go, and find that, a quarter-second later, it is at +3! Evidently the change from 10° is not single-valued – the system is contradicting itself. What are we to do now?

Our difficulty is typical in scientific investigation and is fundamental: we want the transformation to be single-valued but it will not come so. We cannot give up the demand for singleness, for to do so would be to give up the hope of making single-valued predictions. Fortunately, experience has long since shown what is to be done: the system must be redefined.

At this point we must be clear about how a "system" is to be defined. Our first impulse is to point at the pendulum and to say "the system is that thing there". This method, however, has a fundamental disadvantage: every material object contains no less than an infinity of variables and therefore of possible systems. The real pendulum, for instance, has not only length and position; it has also mass, temperature, electric conductivity, crystalline structure, chemical impurities, some radio-activity, velocity, reflecting power, tensile strength, a surface film of moisture, bacterial contamination, an optical absorption, elasticity, shape, specific gravity, and so on and on. Any suggestion that we should study "all" the facts is unrealistic, and actually the attempt is never made. What is necessary is that we should pick out and study the facts that are relevant to some main interest that is already given.

The truth is that in the world around us only certain sets of facts are capable of yielding transformations that are closed and single-valued. The discovery of these sets is sometimes easy, sometimes difficult. The history of science, and even of any single investigation, abounds in examples. Usually the discovery involves the other method for defining of a system, that of listing the variables that are to be taken into account. The system now means not a thing but a list of variables. This list can be varied, and the experimenter's commonest task is that of varying the list ("taking other variables into account") until he finds a set of variables that gives the required singleness. Thus we first considered the pendulum as if it consisted of the variable "angular deviation from the vertical"; we found that the system so defined did not give singleness. If we were to go on we would next try other definitions, for instance the vector:

(1) (*angular deviation, mass of bob*)

which would also be found to fail. Eventually we would try the vector

(2) (*angular deviation, angular velocity*)

and then we would find that these states, defined in this way, would give the desired singleness."

From the physical point of view the phenomenon examined by the experimenter, let us denote it *p*, can be labelled as: "small oscillations of a pendulum with one degree of freedom". This particular pendulum (a particular real dynamic system as W. R. Ashby calls it) on which the experiment is carried out is an instance of that phenomenon. The experimenter would certainly expect that at least all "ideal copies" of the pendulum examined are such instances as well. But any attempt to state more accurate and more complete criteria for distinguishing instances of small oscillation of the pendulum with one degree of freedom among all possible empirical systems must be postponed till more information about the behaviour of various candidates to that class is collected, and till some regularities in their behaviour are grasped by suitable laws (mathematical models).

Observe that W. R. Ashby's remark that the task of the experimenter is that of varying the list of variables until he finds a set "that gives the required singleness" should be completed by adding that the variable "angular deviation" is not treated on a pair with any other. The experimenter's task is not to find any set of variables whatsoever with respect to which the system is deterministic but the one which includes "angular deviation". The reason why the experimenter distingushes some variables need not be of theoretical nature. Some parameters may deserve special attention because of their practical, some others because of their theoretical significance. Note that the customary way of asking questions about empirical systems is "why does the system behave in such and such way?", where the behaviour of the system is described in some variables selected still before any experiment was undertaken.

The fact that some variables are predistinguished makes the task of the experimenter more definite. Still there may exist many different sets of variables which comprise angular deviation and match the requirement of singleness of transformations of states of *p*. Consider for instance the following couple

(3) (*angular deviation, angular deviation k seconds later*)

If the experiment does not last too long so that the changes of the amplitude of oscillations are negligibly small, the experimenter may not be able to detect any violation of the singleness requirement in the behaviour of *p*. What is more (3) has some advantage over (2), namely values of the variable "angular deviation of *p* *k* seconds later" can be measured easier than that of angular velocity. Still from the theoretical point of view the conceptualization (3) is definitely worse than that given by (2). The law which describes the behaviour of *p* in terms of variables in (3) cannot be applied to a pendulum whose length differs from that of *p*. Suppose in turn that instead of oscillations in a fixed plane (i.e. with one degree of freedom) the experimenter wants to study, say, the superposition of two perpendicular oscillations (the oscillation with two degrees of freedom). The conceptualization (2) still works with the proviso that the variables in (2) are treated as vectors while (3) becomes useless.

We see therefore that the expediency of a particular conceptualization of a phenomenon we are interested in cannot be adequately evaluated unless we check how it works as a part of the conceptual framework of a theory which covers as many phenomena close to the examined one as possible. Once again we arrive at the conclusion that the difference between a theory whose scope is sufficiently large and a mathematical model which gives an account for a specific regularity of a specific phenomenon does not consist simply in the fact that their areas of applicability are different. Metaphorically speaking, theories are not sets of models but rather systems of models linked by a common conceptual framework.

IV. MEASUREMENT

4.1. SEMANTIC CONCEPTION OF MEASUREMENT

The exposition of the theory of measurement which the reader will find in this chapter is untypical at least in the following two respects.

The extensional structures which we are going to discuss will not involve concatenation operation, i.e. the operation of combining two or more objects into a new one. Instead, so-called "interval structures" of various sorts will be applied.

Much care will be taken to give a realistic account of fundamentals of measurement. In particular, the notion of approximate (operational) measurement will be discussed in a rather extended way.

A standard and systematic exposition of measurement theory may be found in *Foundations of Measurement*, an excellent monograph by D. H. Krantz, R. D. Luce, P. Suppes, and A. Twerski (1971). Our task will be rather modest. By discussing the concept of measurement we shall try to set up fundamentals for later discussion on the main problem of this book: the nature of relations which the theory bears to phenomena examined. First, however, some technical notions have to be introduced.

DEFINITION 4.1.1. *Let A be a non-empty set and $\preccurlyeq$ be a binary relation on A. We shall say that A is* weakly ordered *by* $\preccurlyeq$, *(*$(A, \preccurlyeq)$ *is a* weakly-ordered set*) if and only if for all a, b, c in A the following conditions are satisfied*:

(R) $a \preccurlyeq a$.

(T) *If* $a \preccurlyeq b$ *and* $b \preccurlyeq c$, *then* $a \preccurlyeq c$.

(C) *Either* $a \preccurlyeq b$ *or* $b \preccurlyeq a$.

i.e. if $\preccurlyeq$ *is reflexive, transitive and connected on A.*

As a matter of fact the reflexivity requirement (R) is redundant since it is an immediate consequence of connectedness (condition (C)) of A.

In what follows we shall write

(1) $a \sim b$, whenever $a \preccurlyeq b$ and $b \preccurlyeq a$,

and

(2) $a \prec b$, whenever $a \preccurlyeq b$ and not $a \preccurlyeq b$.

The relation $\sim$ is easily seen to be an *equivalence* on A (i.e. it is reflexive, symmetric and transitive). We shall occasionally refer to $\sim$ and $\prec$ as to *symmetric* and *asymmetric* parts of $\preccurlyeq$ (Indeed one may easily see that $\prec$ is asymmetric, i.e. $a \prec b$ implies that not $b \prec a$).

The reason why when taking interest in measurement we must as well be interested in ordering relations of various sorts is obvious. When measuring some quantity defined on a set A we try to locate and represent faithfully by means of numbers (or other familiar mathematical entities, such as for instance vectors) the "position" which an object in A occupies relatively to other members of A with respect to certain order displayed by them. The reader should be warned that in these informal comments the term "order" is employed in the loose way in which we use it in the every-day language. Making use of the notion of a weak order is one of the possible ways of transforming the intuitive idea of order into a rigorously defined technical (mathematical) concept.

Continuing our remarks let us notice that, for instance when measuring temperature we try to reflect by means of numbers the relation "is warmer than", when measuring probability we try to reflect the relation "is more likely than", when measuring location on a stright line we try to reflect the relation "is more to the left (to the right) than", etc., etc. The examples of relations quoted above are immediately seen to be examples of the weak-order relations.

The following assertion is trivially valid.

ASSERTION 4.1.2. *Let* **A** = $(A, \preccurlyeq)$ *be a weakly-ordered set, the set* A *being denumerable (i.e. finite or countable). Then there exists a homomorphism* f *from* $(A, \preccurlyeq)$ *into* = $(Re, \leqslant)$, *where* $\leqslant$ *is the familiar "is not greater than" relation defined on Re, i.e. there exists a function* $f\colon A \to Re$ *such that for all* a, b *in* A

(H1) $a \preccurlyeq b$ *if and only if* $f(a) \leqslant f(b)$.

Still more may easily be proved. Namely, one can show that the cardinality of the set of all functions which satisfy the conditions imposed on f is at least as great as that of *Re.* Were those conditions the only ones imposed on f, most likely the numerical representation $(f(A), \leqslant)$ of $(A, \preccurlyeq)$ (regardless of how it were finally selected, i.e. how f were defined) would be of rather restricted use for us. In most cases the fact that A is weakly ordered does not provide all by itself a sufficient base for defining a homomorphism which would enable us to represent the relations among elements of A in a numerical manner.

The following terminological conventions are useful. The homomorphism f is often called a *scale.* The selection of a particular scale to represent $(A, \preccurlyeq)$ may have some practical motivation but from the formal point of view all scales are equally good, they are merely different numerical representations of the same quantity. Observe that a quantity may be conceived to be the class of all scales determined by a weakly ordered set. We shall not, however, try to pursue the issue further and in particular we shall not try to transform these casual remarks into a formal definition. Still we shall make use of the locution "the quantity determined by a weakly ordered set $(A, \preccurlyeq)$" believing that its intuitive content is sufficiently clear.

The prospects of establishing a simple method for determining a numerical representation of a weakly ordered set becomes dramatically improved when apart from the relation $\preccurlyeq$ we are able to define a relation which allows us to compare "differences" between elements of A with respect to $\preccurlyeq$. Let us first discuss the notion of a difference in an informal way.

Assume thus that $(A, \preccurlyeq)$ is a weakly ordered set and select any three objects a, b, c in A. If $a \preccurlyeq b \preccurlyeq c$ then guided by obvious intuitions we may say that the difference between a and b is not greater than that between a and c. We also easily agree that the difference between a and b is the same as that between b and a.

When represented numerically, the difference spoken about becomes a measure of the length of intervals determined by numerical values assigned to objects compared with respect to the weak ordering relation, (the difference between temperatures of two objects, probabilities of two events, or positions of two particles with respect to a selected coordinate.

Let us call any pair (a, b) of elements of A an *interval* in A. If f is a scale of the quantity determined by $(A, \preccurlyeq)$ i.e. for all a, b in A

(H1) $\quad a \preccurlyeq b$ *if and only if* $f(a)$

and the difference between (a, b) is evaluated to be not greater than that between (c, d), we may demand f to satisfy the condition $|f(a)-f(b)| \leqslant |f(c)-f(d)|$. In that way we arrive at certain restricted class of scales for $(A, \preccurlyeq)$ which can be quite easily defined provided that we are able to order intervals with respect to their "length". Denote the comparison relation for intervals as $\underset{\sim}{\sqsubset}$. Then, for all a, b, c, d in A we expect the following condition to be satisfied:

(H2) $\quad (a, b) \underset{\sim}{\sqsubset} (c, d)$ *if and only if* $|f(a)-f(b)| \leqslant |f(c)-f(d)|$.

The following conventions are standard. We shall write $(a, b) \approx (c, d)$ when $(a, b) \underset{\sim}{\sqsubset} (c, d)$ and $(c, d) \underset{\sim}{\sqsubset} (a, b)$, and $(a, b) \sqsubset (c, d)$ when $(a, b) \underset{\sim}{\sqsubset} (c, d)$ but not $(c, d) \underset{\sim}{\sqsubset} (a, b)$, thus introducing notation for symmetric and asymmetric parts of $\underset{\sim}{\sqsubset}$, respectively.

DEFINITION 4.1.3. *Let A be a non empty set, $\preccurlyeq$ a binary relation on A and $\underset{\sim}{\sqsubset}$ a binary relation on $A \times A$. The structure $(A, \preccurlyeq, \underset{\sim}{\sqsubset})$ will be said to be an* interval structure *if and only if for all a, b, c in A the following conditions are satisfied*:

(I1) $(A, \preccurlyeq)$ *is a weakly ordered set.*

(I2) $(A \times A, \underset{\sim}{\sqsubset})$ *is a weakly ordered set.*

(I3) $(a, b) \approx (b, a)$.

(I4) $(a, a) \underset{\sim}{\sqsubset} (b, c)$.

(I5) *If* $a \sim c, a \preccurlyeq b, c \preccurlyeq d$, *then*
$(a, b) \sqsubsetsim (c, d)$ *if and only if* $b \preccurlyeq d$,
and
$(c, d) \sqsubsetsim (a, b)$ *if and only if* $d \preccurlyeq b$.

(I6) *If* $b \sim d, a \preccurlyeq b, c \preccurlyeq d$, *then*
$(a, b) \sqsubsetsim (c, d)$ *if and only if* $c \preccurlyeq a$,
and
$(c, d) \sqsubsetsim (a, b)$ *if and only if* $a \preccurlyeq c$.

(I7) *If* $(a_1, b_1) \sqsubsetsim (a_2, b_2)$, $(b_1, c_1) \sqsubsetsim (b_2, c_2)$, $a_1 \preccurlyeq b_1 \preccurlyeq c_1$ *and* $a_2 \preccurlyeq b_2 \preccurlyeq c_2$, *then* $(a_1, c_1) \sqsubsetsim (a_2, c_2)$.

Conditions (I1) – (I7) imposed on $\preccurlyeq$ and $\sqsubsetsim$ are easily seen to be necessary in order for a function f that satisfies (H1), (H2) to exist. By producing a suitable example one may prove that they do not suffice for that purpose. Perhaps it would be worthwhile examining whether any finite set of conditions stated in an elementary language suffices to guarantee the existence of such a function. Personally I think it very unlikely. But, although interesting from the formal point of view, the issue is irrelevant for the discussion we are going to undertake.

Let me notice also that conditions (I1) – (I7) are not independent, for instance (I1) can be derived from the remaining ones, a similar remark applies to (I2). But in the form in which it has been stated, Definition 4.1.3. is probably more transparent than if it were based on independent conditions, and it allows to grasp substantial properties of interval structures more easily.

ASSERTION 4.1.4. *If* $\mathbf{A} = (A, \preccurlyeq, \sqsubsetsim)$ *is an interval structure, then* $\sim$ *is a congruence in* $\mathbf{A}$, *i.e. for each* $a, b, c, d, a', b', c', d'$ *in* A *if*

$$a \sim a', b \sim b', c \sim c', d \sim d',$$

then

(C1) $a \preccurlyeq b$ *if and only if* $a' \preccurlyeq b'$,

(C2) $(a, b) \sqsubsetsim (c, d)$ *if and only if* $(a', b') \sqsubsetsim (c', d')$.

PROOF. Assume that $a \preccurlyeq b$. Then $a' \preccurlyeq a \preccurlyeq b \preccurlyeq b'$ and hence $a' \preccurlyeq b'$. Clearly $a' \preccurlyeq b'$ implies $a \preccurlyeq b$ in the same way. Thus we have established (C1).

In turn assume that $(a, b) \sqsubsetsim (c, d)$ and suppose that $a \preccurlyeq b$.

We have $a \preccurlyeq b'$. Substitute in (I5) a for c, and b' for d, then (I5) yields: $(a, b) \sqsubseteq (a, b')$ if and only if $b \preccurlyeq b'$ and $(a, b') \sqsubseteq (a, b)$ if and only if $b' \preccurlyeq b$. Since $b \sim b'$ we obtain $(a, b) \approx (a, b')$. Applying (I6) one may prove by a similar argument that $(a, b) \approx (a', b')$. Thus by (I2), which guarantees transitivity of $\approx$, we obtain $(a, b) \approx (a', b')$. Clearly, we have also $(c, d) \approx (c', d')$. Then $(a', b') \approx (a, b) \approx (c, d) \approx (c', d')$, and hence $(a', b') \approx (c', d')$. To prove the "only if" part of (C2) we simply have to reverse the argument.

ASSERTION 4.1.5. *If* $(A, \preccurlyeq, \sqsubseteq)$ *is an interval structure, then the following conditions are valid for all* a, b, c, d *in* A:

(a) $a \sim b$ *if and only if* $(a, b) \sqsubseteq (a, a)$.

(b) $(a, a) \approx (b, b)$.

(c) *If* $a \sim c, a \preccurlyeq b, c \preccurlyeq d$, *then*

$$(a, b) \sqsubset (c, d) \text{ if and only if } b \prec d,$$

and

$$(a, b) \approx (c, d) \text{ if and only if } b \sim d.$$

(d) *If* $b \sim d$, $a \preccurlyeq b$, $c \preccurlyeq d$, *then*

$$(a, b) \sqsubset (c, d) \text{ if and only if } c \prec a,$$

and

$$(a, b) \approx (c, d) \text{ if and only if } c \sim a.$$

(e) *If* $a \preccurlyeq b \preccurlyeq c$, *then* $(a, b) \sqsubseteq (a, c)$ *and* $(b, c) \sqsubseteq (a, c)$.

(f) *If* $a \prec b \prec c$, *then* $(a, b) \sqsubset (a, c)$ *and* $(b, c) \sqsubset (a, c)$.

(g) *If* $(a_2, b_2) \sqsubseteq (a_1, b_1)$, $(b_2, c_2) \sqsubseteq (b_1, c_1)$, $a_1 \preccurlyeq b_1 \preccurlyeq c_1$, *and* $a_2 \preccurlyeq b_2 \preccurlyeq c_2$, *then* $(a_2, c_2) \sqsubseteq (a_1, c_1)$.

(h) *If* $(a_1, b_1) \approx (a_2, b_2)$, $(b_1, c_1) \approx (b_2, c_2)$, $a_1 \preccurlyeq b_1 \preccurlyeq c_1$, *and* $a_2 \preccurlyeq b_2 \preccurlyeq c_2$, *then* $(a_1, c_1) \approx (a_2, c_2)$.

PROOF.

(a) By (C2) in Assertion 4.1.4. we obtain the implication from left to right. Assume that $(a, b) \sqsubseteq (a, a)$, and put $c = d = a$ in (I5). Since $a \sim a$ and $a \preccurlyeq a$ we conclude that: if $a \preccurlyeq b$, then $(a, b) \sqsubseteq (a, a)$ if and only if $b \preccurlyeq a$. By the assumption we have made, we get: if $a \preccurlyeq b$ then $b \preccurlyeq a$, or equivalently: if $a \preccurlyeq b$ then $a \sim b$. Suppose, however, that $b \prec a$. In this case we have to rewrite our assumption in the form $(b, a) \sqsubseteq (a, a)$ which, by (I3), is equivalent to the original one. Applying the same argument as before we arrive at the conclusion that if $b \preccurlyeq a$ then $a \sim b$. But, by connectedness of $\preccurlyeq$ we have $a \preccurlyeq b$ or $b \preccurlyeq a$, and thus in either case $a \sim b$.

(b) By (I4) we have both $(a, a) \sqsubseteq (b, b)$ and $(b, b) \sqsubseteq (a, a)$, and thus $(a, a) \approx (b, b)$.

(c) Suppose that the assumptions of (c) are satisfied and $b \preccurlyeq d$. Since by (I5) $(c, d) \sqsubseteq (a, b)$ entails $d \preccurlyeq b$, we obtain $(c, d) \sqsubset (a, b)$. In turn assume that $(a, b) \sqsubset (c, d)$. But then again by (I5), $d \preccurlyeq b$ would contradict this assumption, hence $b \prec d$. In this way we have established the first part of (c). The second part is an immediate corollary from (I5).

(d) Apply the same argument as in case (c), but use (I6) instead of (I5).

(e) Assume that $a \preccurlyeq b \preccurlyeq c$ and in (I5) write everywhere a instead of c and c instead of d. One of the resulting formulas will be: if $a \sim a$, $a \preccurlyeq b$, $a \preccurlyeq c$, then $(a, b) \sqsubseteq (a, c)$ if and only if $b \preccurlyeq c$. This gives $(a, b) \sqsubseteq (a, c)$. In turn apply (I6) in a similar manner. Write c instead of b and d, replace a by b, and c by a. You will obtain: if $c \sim c$, $b \preccurlyeq c$, $a \preccurlyeq c$, then $(b, c) \sqsubseteq (a, c)$ if and only if $a \preccurlyeq b$. This gives $(b, c) \sqsubseteq (a, c)$ and concludes the proof.

(f) Apply an argument similar to that used in case (e) but based on (c) and (d) instead of (I5) and (I6).

(g) Substitute in (I7) a_1, b_1, c_1 for all occurrences of a_2, b_2, c_2 and at the same time substitute a_2, b_2, c_2 for all occurrences of a_1, b_1, c_1 respectively. The resulting formula will be (g).

(h) By (I7) and (g).

The notion we are going to define now will play the key role in the considerations carried out in this chapter.

DEFINITION 4.1.6. *A structure* $(A, \preccurlyeq, \sqsubseteq)$ *is a* (interval) measurement structure *if and only if the following conditions are satisfied*:

(M1) $(A, \preccurlyeq, \sqsubseteq)$ *is an interval structure.*

(M2) *If* $c \preccurlyeq d$, *and* $(a, b) \sqsubseteq (c, d)$, *then there exist* a', b', *such that* $c \preccurlyeq a', b' \preccurlyeq d$ *and* $(a, b) \approx (c, b') \approx (a', d)$,

(M3) *If* $a \prec b$ *and* $(a, b) \sqsubseteq (c, d)$, *then there exists a finite sequence* $b_1, \ldots, b_n$ *of elements of* A *such that*:

(i) $a \prec b_1 \prec b_2 \prec \ldots \prec b_n$,
(ii) $b = b_1$,
(iii) $(a, b_1) \approx (b_1, b_2) \approx \ldots \approx (b_{n-1}, b_n)$,
(iv) $(c, d) \sqsubset (a, b_n)$, and
(v) *if* $(c, d) \underset{\approx}{\sqsubset} (a, b_b)$, *then* $(a, b_{n-1}) \sqsubset (c, d)$.

In what follows each sequence which satisfies clauses (i) – (v) of (M3) will be called a *standard sequence*. We shall prove that:

LEMMA 4.1.7. *For all* a, b, c, d *which satisfy the assumptions of condition* (M3) *if* $b'_1, \ldots, b'_n$ *and* $b''_1, \ldots, b''_m$ *are standard sequences satisfying* (i) – (v), *then* $n = m$.

PROOF. Observe that $b'_1 = b''_1 = b$, and thus we immediately have $(a, b'_1) \approx (a, b''_1)$. By (iii) we know that $(a, b'_1) \approx (b'_1, b'_2) \approx \ldots \approx (b'_{n-1}, b'_n)$ and similarly $(a, b''_1) \approx (b''_1, b''_2) \approx \ldots \approx (b''_{n-2}, b''_m)$ The transitivity of $\approx$ yields that

(1) $$(a, b'_1) \approx (b_i, b_{i+1}) \approx (b''_j, b''_{j+1}) \approx (a, b''_1),$$

for all $i = 1, \ldots, n-1$, $j+1, \ldots, m-1$.

Since $a \prec b'_1 \prec \ldots \prec b'_n$ and $a \prec b''_1 \prec \ldots \prec b''_n$ we may apply (1) and Assertion 4.1.5 (h) in order to prove that for each $k = 1, \ldots, min(n, m)$

(2) $$(a, b'_k) \approx (a, b''_k)$$

As a matter of fact, if $k = 1$, (2) becomes a part of (1). If $k = 2$ the argument is a trivial application of Assertion 4.1.5 (h). Having established (2) for $k = 2$, in order to establish it for $k = 3$ one must apply the assertion once again. Repeting the procedure one eventually arrives at $k = min(n, m)$.

Without any loss of generality we may assume that $n \leqslant m$. If $(a, b'_n) \approx (c, d)$, then by (2) and transitivity of $\approx$ we obtain

(3) $$(a, b''_n) \approx (c, d)$$

which proves that $n = m$. Indeed, suppose that apart from (3),

(4) $$(a, b''_m) \approx (c, d)$$

for some $m > n$. From (3) and (4) by transitivity of $\approx$ we obtain

(5) $$(a, b''_n) \approx (a, b''_m).$$

But $a \prec b''_1 \prec \ldots \prec b''_n \prec \ldots \prec b''_m$ and thus in particular $a \prec b''_n \prec b''_m$. By 4.1.5 (f) we obtain $(a, b''_n) \sqsubset (a, b''_m)$ which contradicts (5).

In turn assume that

(6) $(a, b'_{n-1}) \sqsubset (c, d) \sqsubset (a, b'_n)$.

Then again (2) yields

(7) $(a, b''_{n-1}) \sqsubset (c, d) \sqsubset (a, b''_n)$,

and again a simple argument similar to that we have applied when comparing (3) and (4) allows to establish that $n = m$, which concludes the proof.

THEOREM 4.1.8. *If $(A, \preccurlyeq, \sqsubsetapprox)$ is a measurement structure, then for each e_0, e_1, e_2 in A, and for each ξ_0, δ in Re if not $e_1 \sim e_2$ and $\delta > 0$, then there exists exactly one mapping f: $A \to Re$ such that the following conditions are satisfied*:

(i) $f(e_0) = \xi_0$,

(ii) $|f(e_1) - f(e_2)| = \delta$,

(iii) *Let L be a binary relation on $Re \times Re$ defined as follows. For all x, y, x', z', $(x, y)\ L(x', z')$ if and only if $|x-y| \leqslant |x'-y'|$. Then f is a homomorphism from $(A, \preccurlyeq, \sqsubsetapprox)$ into $(Re, \leqslant, L)$, i.e. for all a, b, c, d in A the following two familiar conditions are satisfied*:

(H1) *$a \preccurlyeq b$ if and only if $f(a) \leqslant f(b)$.*

(H2) *$(a, b) \sqsubsetapprox (c, d)$ if and only if $|f(a) - f(b)| \leqslant |f(c) - f(d)|$.*

In what follows the structure $(Re, \leqslant, L)$ defined above will be dealt with very often. Let us then, in order to abbreviate the notation denote it as **Re**.

Before we start proving the theorem let us notice that the following is valid.

ASSERTION 4.1.9. *If f is a homomorphism from $(A, \preccurlyeq, \sqsubsetapprox)$ into $(Re, \leqslant, L)$, then for all a, b, c, d in A*:

(H3) *$a \prec b$ if and only if $f(a) \leqslant f(b)$.*

(H4) *$a \sim b$ if and only if $f(a) = f(b)$.*

(H5) *$(a, b) \sqsubset (c, d)$ if and only if $|f(a) - f(b)| < |f(c) - f(d)|$.*

(H6) *$(a, b) \approx (c, d)$ if and only if $|f(a) - f(b)| = |f(c) - f(a)|$.*

PROOF. Suppose that $a \prec b$ and $f(a) = f(b)$. By (H1) we obtain $b \preccurlyeq a$, which is impossible. This proves that if $a \prec b$, then $f(a) < f(b)$. In turn assume that $f(a) < f(b)$ and $a = b$. This time, (H1) yields $f(b) \leqslant f(a)$, which again is impossible, and hence $f(a) < f(b)$ entails $a \prec b$. This concludes the proof of (H3). The proof of (H5) can be obtained by an analogous argument. (H4) is an immediate consequence of (H1) and (H3), and (H6) is an immediate consequence of (H2) and (H5).

Now we can turn back to Theorem 4.1.8.

Proof of Theorem 4.1.8. Assume that e_0, e_1 and e_2 are fixed. Let $e_1 \prec e_2$. The proof does not depend on a particular selection of x_0 and δ and we may put, for instance, $f(a_0) = 0$, $|f(e_1) - f(e_2)| = 1$. The decision amounts to selecting the origin and the unit of the scale applied while measuring the quantity represented by f (or equivalently: the quantity determined by $\preccurlyeq$ and $\sqsubsetapprox$). Thus for instance the transformation from Celsius' to Reumir's scale consists in changing the unit, from Celsius' to Kelvin's in changing the zero point (the origin of the scale), and finally from Celsius' to Fahrenheit's in changing both. Due to some practical considerations wewmay be inclined to reckon some of the choices of the origin and unit to be better than some other ones, but clearly nothing empirical compels any of those choices.

Single out any a in A. We shall subsequently discuss several separate cases.

Case 0. $a \sim e_0$. In that case in view of (H4) we must put $f(a) = f(e_0) = 0$.

Case A. $e_0 \prec a$. We certainly must put $f(a) > 0$, but this is only a preliminary condition imposed on the value of $f(a)$. Compare the intervals (e_0, a) and (e_1, e_2). If, by any chance, $(e_0, a) \approx (e_1, e_2)$, then clearly in order for f to satisfy (H6) we must put $|f(e_0) - f(a)| = |f(e_1) - f(e_2)| = 1$. Since $f(e_0) = 0$ we obtain $f(a) = 1$, thus determining $f(a)$ in a unique way. But it may also happen that:

Case A.1. $(e_0, a) \sqsubset (e_1, e_2)$. The assumptions of condition (M3) in Definition 4.1.2. are satisfied, and we may form a sequence

$a_1, \ldots, a_n$ such as defined in (M3), i.e. a standard one. The simplest possibility which may take place is

(1) $$(e_0, a_n) \approx (e_1, e_2).$$

Since

(2) $$(e_0, a_1) \approx (a_1, a_2) \approx \ldots \approx (a_{n-1}, a_n),$$

and $a = a_1$, we must assign values to $f(a_i)$ in such a way that $|f(a_{i-1}) - f(a_i)| = |f(e_0) - f(a_1)| = |f(a)| = f(a)$. Otherwise we would violate (H6). On the other hand $e_0 \prec a_1 \prec \ldots \prec a_n$, therefore in order for (H3) to be satisfied we must put $f(e_0) < f(a_1) < \ldots < f(a_n)$. We have then

(3) $$f(a_k) = kf(a),$$

for all $k = 1, \ldots, n$. But $f(a_n) = |f(e_0) - f(a_n)|$, which by (1) and (H6) yields

(4) $$f(a_n) = 1.$$

As an immediate corollary from (3) and (4) we have

(5) $$f(a) = \frac{1}{n}.$$

By Lemma 4.1.7 we know that n does not depend on a particular selection of the standard sequence, therefore the value of $f(a)$ is determined in a unique way.

Suppose however that (1) does not hold true. Then by clause (v) of (M3) we obtain:

Case A.1.1. $(e_0, a_{n-1}) \sqsubset (e_1, e_2) \sqsubset (e_0, a_n)$. It follows from (M2) then that there must exist an object a'_{n-1} such that

(6) $$(e_0, a_{n-1}) \approx (e_1, a'_{n-1}),$$

and similarly there must exist an object e'_2 such that

(7) $$(e_0, e'_2) \approx (e_1, e_2).$$

The following diagram should help in figuring out the situation.

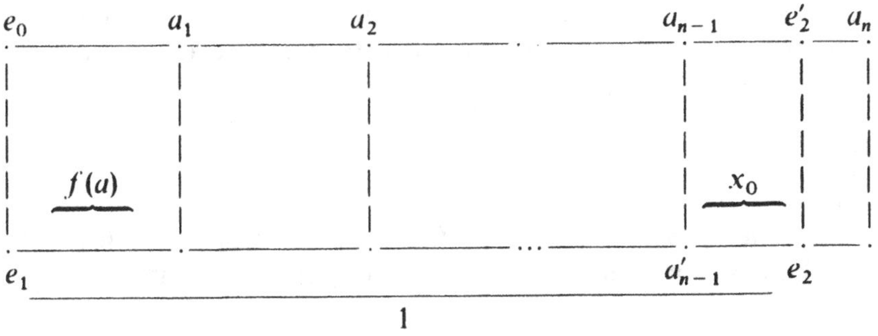

Suppose that we have determined the value of $|f(a_{n-1}')-f(e_2)|$. Let it be x_0. We have $|f(e_1)-f(a_{n-1}')|+|f(a_{n-1}')-f(e_2)| = |f(e_1)-f(e_2)| = 1$. Thus $|f(e_1)-f(a_{n-1}')| = 1-x_0$. By equivalence (6) we obtain $|f(e_0)-f(a_{n-1})| = f(a_{n-1}) = 1-x_0$, which by (3) yields

$$(8) \qquad f(a) = \frac{1-x_0}{n-1}.$$

The procedure by means of which one may try to evaluate the value of x_0 is exactly the same as that we have applied trying to evaluate $f(a)$. Instead of comparing $|f(e_0)-f(a_1)|$ and $|f(e_1)-f(e_2)|$ we must now compare $|f(a_{n-1}')-f(e_2)|$ and $|f(e_1)-f(e_2)|$. Since $(a_{n-1}', e_2) \approx (a_{n-1}, e_2') \approx (a_{n-1}, a_n) \sqsubset (e_0, a_1)$, we have $|f\times \times(a_{n-1}')-f(e_2)| < |f(e_0)-f(a_1)|$, i.e. $x_0 < f(a)$.

There are two possibilities. Either, for some $n_1 > n$, $x_0 = \dfrac{1}{n_1}$ or for some $x_1 < x_0$, $x_0 = \dfrac{1-x_1}{n_1-1}$. In the latter case we must repeat the whole procedure once again. If subsequent repetitions eventually end in establishing the value of some x_k, we shall find $f(a)$ by solving $k+1$ equations

$$(9) \qquad f(a) = \frac{1-x_0}{n-1}, \quad x_0 = \frac{1-x_1}{n_1-1}, \ldots, \quad x_k = \frac{1-x_k}{n_k-1}$$

in k unknown.

Suppose however that we can never arrive at the desired end, i.e.

Case A. 1.2. the sequence $x_0, x_1, \ldots, x_i, \ldots$ is infinite. Compare the values of x_0 and $f(a)$. We have already established that

$x_0 < f(a)$. Now we shall show that $n_1 > n$. To this aim let us form the standard sequence in order to compare (a_{n-1}, e'_2) and (e_1, e_2). Let it be $b_1, \ldots, b_{n1}$. We have $b_1 = e'_2$,

(10) $$(a_{n-1}, b_1) \approx (b_1, b_2) \approx \ldots \approx (b_{n1-1}, b_{n1}),$$

and $(a_{n-1}, b_{n1-1}) \sqsubset (e_1, e_2)$.

Since $(a_{n-1}, b_1) = (a_{n-1}, e'_2)$, $(a_{n-1}, b_1) \sqsubset (a_{n-1}, a_n) \approx (e_0, a_1)$, and therefore, by transitivity of $\sqsubsetapprox$ and (10), we obtain

(11) $$(a_{n-1}, b_{n-1}) \sqsubset (e_0, a_{n-1}).$$

Observe now that (10) implies that in particular

(12) $$(b_{n-1}, b_n) \approx (a_{n-1}, e'_2).$$

We have also

(13) $$a_{n-1} \preccurlyeq b_{n-1} \preccurlyeq b_n,$$

and

(14) $$e_0 \preccurlyeq a_{n-1} \preccurlyeq e'_2.$$

By applying condition (I7) of Definition 4.1.3 to (11), (12), (13), and (14) we obtain

(15) $$(a_{n-1}, b_n) \sqsubsetapprox (e_0, e'_2),$$

but $(e_0, e'_2) \approx (e_1, e_2)$ and therefore if $(a_{n-1}, b_n) \approx (e_0, e'_2)$, then the value of x_0 would be determined, namely we would have $x_0 = 1/n$ contrary to the assumption that defines case A. 1.2 being discussed now. We have then

(16) $$(a_{n-1}, b_n) \sqsubset (e_0, e'_2) \approx (e_1, e_2)$$

and therefore $n_1 \geqslant n+1$.

Let us examine how the fact that $n_1 \geqslant n+1$, and thus in general $n_i \geqslant n_{i-1}+1$, $i = 2, 3, \ldots$, can help in determining the value of $f(a)$. Observe that since

(17) $$(e_0, a_{n-1}) \sqsubset (e_1, e_2) \sqsubset (e_0, a_n),$$

therefore

(18) $$f(a_{n-1}) < 1 < f(a_n).$$

The latter inequalities are by (3) equivalent to

(19) $$\frac{1}{n} < f(a) < \frac{1}{n-1}.$$

The analogon of (19), which must be valid for all x_i, is

(20) $$\frac{1}{n_{i+1}} < x_i < \frac{1}{n_{i-1}-1}.$$

Since $n < n_1 < n_2 < \ldots$, i.e. n_i form an increasing sequence,

(21) $$\lim_{k\to\infty} x_k = 0.$$

On the other hand we have by (9)

(22) $$f(a) = \frac{1}{n-1} - \frac{x_0}{n-1} = \frac{1}{n-1} - \frac{1}{(n-1)(n_1-1)} - \frac{x_1}{(n-1)(n_1-1)} = \frac{1}{n-1} - \frac{1}{(n-1)(n_1-1)} - \frac{1}{(n-1)(n_1-1)(n_2-1)} - \frac{x_2}{(n-1)(n_1-1)(n_2-1)} = \ldots$$

and since the sum

(23) $$\sum_{k=1}^{\infty}\left(\frac{1}{(n-1)(n_1-1)\ldots(n_k-1)}\right) \leqslant \sum_{k=1}^{\infty}\frac{1}{k!}$$

is convergent, we have

(24) $$f(a) = \lim_{k\to\infty}\left(\frac{1}{n-1} - \sum_{i=1}^{k}\frac{1}{(n-1)(n_1-1)\ldots(n_k-1)}\right).$$

The next case we have to discuss is

Case A.2. $(e_1, e_2) \precnapprox (e_0, a)$. We do not face here any new problem. Instead of constructing the standard sequence whose initial element is e_0 and whose neighborhood elements form intervals equivalent to (e_0, a), one must now construct the standard sequence with the same initial element but with intervals between neighborhood elements equivalent to (e_1, e_2). While the procedure described in *Case A* may be said to be that of evaluating the difference between e_1 and e_2 by means of the interval (e_0, a), the procedure one has to apply now is the reverse one: the difference between e_0 and e must be evaluated by means of the interval (e_1, e_2).

Finally let us briefly comment on

Case B. $a \prec e_0$. Obviously we must put $f(a) < f(e_0) = 0$, i.e. assign to a a negative real number. But the absolute value $|f(a)|$ of $f(a)$ should be established exactly in the same way as in *Case A*, i.e. by comparing what may be called the relative length of the intervals (a, e_0), (e_1, e_2). One must once more apply the technique of constructing an appropriate standard sequence. What is new is that this time the initial point of such a sequence must be a not e_0. But the details of the procedure are exactly the same as in *Case A*.

We have shown that the conditions imposed on f by the assumptions of Theorem 4.1.8 determine f uniquely. In order to complete the proof one must verify that f satisfies both (H1) and (H2), i.e. it is a homomorphism from the measurement structure $(A, \preccurlyeq, \precsim)$ into **Re**. This part of the proof does not involve any essentially new types of argument and will be left to the reader.

COROLLARY 4.1.10. *If* f_1, f_2 *are homomorphisms from a measurement structure* $(A, \preccurlyeq, \precsim)$ *into* **Re**, *then there are such* α *and* β *that for each* a *in* A, $f_1(a) = \alpha + \beta f_2(a)$.

Of some importance is also the following

COROLLARY 4.1.11. *If* A *is a non-empty set,* $\preccurlyeq$ *is a binary relation on* A, *and* $\precsim$ *is a binary relation on* $A \times A$, *then* $(A, \preccurlyeq, \precsim)$ *is a measurement structure if and only if there exists a homomorphism* f *from* $(A, \preccurlyeq, \precsim)$ *into* **Re** *and* $(f(A), \leqslant, L)$ *(i.e. the restriction of* **Re** *to* $f(A)$) *is a measurement structure.*

PROOF. The case when A is a unit set is trivial. One may easily verify that each structure of the form $(\{a\}, \preccurlyeq, \precsim)$ is a measurement structure and each assignment of any number to a is a homomorphism from that structure into **Re**.

If A involves at least two objects then the assumption that $(A, \preccurlyeq, \sqsubseteq)$ is a measurement structure yields the existence of a homomorphism of required sort in virtue of Theorem 1.4.8. The reader may easily verify that the structure $(f(A), \leqslant, L)$ defined by f satisfies conditions (M1) – (M3) of Definition 4.1.5, i.e. it is a measurement structure.

Suppose in turn that f is such a homomorphism. Then conditions (H1), (H2) allow us to prove that $(A, \preccurlyeq, \sqsubseteq)$ must be an interval structure. In order to prove existence of the objects mentioned in hypotheses (M2) and (M3) one should exploit (in an obvious way) the assumption that $(f(A), \leqslant, L)$ is a measurement structure.

Before we comment on both the theorem and the notion of an interval measurement structure let us discuss a few examples which should account for how the theorem and the concept it refers to work. But this will be the objective of the next section.

4.2. INTERVAL MEASUREMENT STRUCTURES

Let us start our list of examples of different measurement structures with discussing the procedure of counting objects.

EXAMPLE 1. Let X be a set of any physical objects. Counting consists in assigning numbers to subsets of X. The measurement structure pertinent to the procedure should then be of the form

$$\mathbf{C} = (2^X, \preccurlyeq, \sqsubseteq).$$

The relation $\preccurlyeq$ is that of "being not more numerous than". The operational meaning of it is simple. In order to compare two elements of 2^X, i.e. two subsets A, B of X with respect to $\preccurlyeq$, one has to try to assign to each object from A exactly one object from B. Different elements of A should be assigned to different elements of B. If that can be done, A is at most as numerous as B is, i.e. either $A \preccurlyeq B$ or $A \sim B$. If not, i.e. if the attempt to carry out the assignment stops before the task has been accomplished because B turns out to be not numerous enough, then A does not bear the relation $\preccurlyeq$ to B, but clearly the procedure results in establishing that $B \preccurlyeq A$.

The way in which the operation $\precsim$ should be understood is obvious. When two couples (A, B), (C, D) of subsets of X are given, we should couple elements of A with those of B and elements of C with those of D in exactly the same way as it should be done when deciding whether the relation $\preccurlyeq$ holds or not between a given pair of sets. Let E_1 be the surplus of elements of either A or B, i.e. the set of elements left uncoupled after the operation of coupling the elements of A with those of B has been ended, and let E_2 be the surplus left after coupling the elements of C and D. We define $(A, B) \precsim (C, D)$ when $E_1 \preccurlyeq E_2$, and $(C, D) \precsim (A, B)$ when $E_2 \preccurlyeq E_1$.

The relation $\preccurlyeq$ we have tried to define above in an operational manner possesses a transparent set-theoretical meaning. The same can be said about $\precsim$. Let us call structures of the discussed sort counting structures, *CN-structures* for short. It is not a difficult task to provide a set theoretical definition of a counting structure, thus setting up a mathematical model for it. This can enlarge our list of examples of how mathematics enters into factual theories becoming eventually unseparable constituent of them.

Observe that in order for $\preccurlyeq$ and $\precsim$ to be defined we should either assume that X is finite or restrict 2^X to $Fin(X)$ (the set of finite subsets of X). Incidentally this remark can be motivated either by examining, at least implicitly, the properties of mathematical models of CN-structures or by pointing out that for infinite sets neither the relation $\preccurlyeq$ nor $\precsim$ has any operational meaning. In fact the latter motivation involves some amount of set-theoretical theorizing since from the operational point of view the notion of an infinite set must be meaningless. There are no operational criteria of the infiniteness of sets.

In order to conform to the remark above let X be assumed to be finite. Is **C** a measurent structure? Does it provide necessary basis for describing in quantitative terms the way in which elements of 2^X are ordered by $\preccurlyeq$ and $\precsim$? As long as we insist on characterizing the structure only in the operational way, the question cannot be provided with a conclusive negative answer unless we show by a suitable experiment that the attempt to define the quantity "the number of elements of A", denote it as $n(A)$, as a homomorphism from **C** into **Re** leads to inconsistency.

To make it entirely plain let us note that if **C** is meant to be a structure defined in the operational way one cannot prove that the structure is a measurement structure, one may, at best, prove that it is not. As a matter of fact the situation is typical of the experimental sciences; the notions they involve are as a rule constructed on the basis of certain assumptions which are subject to empirical verifications.

The way in which one may try to use the structure **C** in order to determine the values of n is obvious. As usual at the very beginning three objects A_0, B_0, C_0 in the structure must be selected so that B_0 does not bear the relation $\sim$ to C_0. Next the conditions $n(A) = \xi_0$ and $|n(B_0)-n(A_0)| = \delta > 0$ must be imposed on n, where ξ_0 and δ are selected in an arbitrary manner. When this is done, i.e. when both the origin and the unit are set up, we are ready to start determining the values of n by implementing the procedures described in the proof of Theorem 4.1.8.

For example suppose that $X = \{a_1, ..., a_{10}\}$. Put $n(\varnothing) = 0$ $|n(\{a_1\})-n(\varnothing)| = 1$. It depends on the results of an experiment whether **C** will turn out to be a measurement structure or not, assume, however, that the experiments fully coincide with what one may expect under the obvious set-theoretical interpretation of $\preccurlyeq$ and $\precsim$. How should one proceed in order to find out what the value of, say, $n(\{a_3, a_4, a_5\})$ is?

After establishing that

$$(\varnothing, \{a_1\}) \precsim (\varnothing, \{a_3, a_4, a_5\}),$$

we must form a standard sequence. One may easily verify that the sequence

$$\{a_1\}, \{a_1, a_2\}, \{a_1, a_2, a_3\}$$

satisfies the conditions imposed on standard sequences, i.e.

(1) $$\{a_1\} \prec \{a_1, a_2\} \prec \{a_1, a_2, a_3\}$$

and

(2) $(\varnothing, \{a_1\}) \approx (\{a_1\}, \{a_1, a_2\}) \approx (\{a_1, a_2\}, \{a_1, a_2, a_3\}).$

On the other hand

(3) $\{a_1, a_2, a_3\} \approx \{a_3, a_4, a_5\}$

and therefore we must put

(4) $n(\{a_1, a_2, a_3\}) = n(\{a_3, a_4, a_5\})$

or otherwise n would not be a homomorphism from **C** into **Re**.

Equivalences (2) give

(5) $$|n(\varnothing)-n(\{a_1\})| = |n(\{a_1\})-n(\{a_1, a_2\})| = = |n(\{a_1, a_2\})-n(\{a_1, a_2, a_3\})|.$$

By the assumption that we have made

(6) $|n(\varnothing)-n(\{a_1\})| = 1,$

and since $n(\varnothing) = 0$ and $\varnothing \preccurlyeq \{a_1\}$, we obtain $n(\{a_1\}) = 1$. In turn by (5), (6) we obtain

(7) $|n(\{a_1\})-n(\{a_1, a_2\})| = |1-n(\{a_1, a_2\})| = 1$

and since $\{a_1\} \prec \{a_1, a_2\}$, we must put $n(\{a_1, a_2\}) = 2$. Proceeding further with those trivial calculations we shall finally establish that

$$|n(\{a_3, a_4, a_5\})| = |n(\{a_1, a_2, a_3\})| = 3$$

which ends "measuring" of $n(\{a_1, a_2, a_3\})$.

The reason why *measuring* has been taken in quotation marks should be obvious. Simply, we are not accustomed to applying the word "measure" to the procedure of counting objects. But from a more fundamental point of view, however, there is no substantial difference between counting objects and, say, measuring electric charge.

Let us figure out that it turns out that there is no homomorphism from **C** into **Re**. Suppose, for example, that for some A, B, C, in 2^X, $A \preccurlyeq B$, $B \preccurlyeq C$ but $C \prec A$, thus the relation $\preccurlyeq$ is not transitive.

There are two possible reactions to the situation described: to revise operational procedures by means of which **C** has been defined or to give up the idea of applying the structure **C** as a measurement structure. The notion of the number of objects of a set seems to be so fundamental that we may hardly imagine the situation in which it would not be applicable. But nowadays the science has reached the stage at which we are wise enough not to reject any possibility only because it seems unconceivable.

If it happened that for some "crazy" part of the world, represented by a set 2^X, it is impossible to define the relations $\preccurlyeq$, $\precsim$ in the way which both conforms to their intuitive meanings established in dealing with numerous CN-structures and matches formal conditions imposed on all measurement structures, then we would have to consider seriously the possibility to give up the idea of applying the notion of a number to the subsets of 2^X (perhaps only some of them) and to start learning to live somehow without that.

Let us examine the next example, perhaps a bit more sophisticated than the one we have just discussed.

EXAMPLE 2. The definition of the relation $\preccurlyeq$ by means of which masses of different objects may be compared depends on objects we want to deal with. For solid bodies not too large and not too small in size, $\preccurlyeq$ may be defined, in an operational manner, with the help of an equal arm balance. The friction of the balance should be sufficiently small in comparison to the weight of the objects whose masses are to be compared, otherwise the experiments would most likely bring results of no use, i.e. escaping any reasonable theoretical interpretation.

The remarks made above, although trivial from the point of view of the physicist, are in a rather drastic disagreement with the radically operational point of view. The independence of operational definitions of any theoretical considerations seems to be one of the fundamental assumptions of operationalism. I decided to mention this divergence between the physicist's and philosopher's way of viewing the process of designing an experiment in order to emphasize that by the very fact of using locutions which involve the objective "operational" we do not neccessarily have to adhere to all dogmas of the operationalists' doctrine.

Let me call attention to another important point. It is customary and convenient to think about mass as a universal property, i.e. the one that may be attributed to any physical object whatsoever. Regardless of whether this tenet is adequate or not, it does not justify all by itself the belief that there exists a universal way of measuring mass. Whatever practical method of mass measuring might be selected, there is only relatively narrow class of objects whose mass can be measured with that method. In particular, whatever balance would be taken as a measuring instrument, its size, the friction of its parts, not to mention other possible parameters which may affect the procedure of measurement, restrict drastically the set of objects whose masses can be compared with the help of the balance.

Let us then agree that, from the operational point of view, there exists no universal measurement structure $\mathbf{U} = (U, \preccurlyeq, \lesssim)$, where U is the set of all physical objects, $\preccurlyeq$ and $\lesssim$ are suitable comparison relations defined in the operational manner which has underlain the definition of mass. Rather, we have a sequence $\mathbf{U}_i = (U_i, \preccurlyeq_i, \lesssim_i)$, $i = 1, 2, \ldots$, of structures whose elements form an ordered set in the sense that for any two structures $\mathbf{U}_i, \mathbf{U}_j$ there exists a structure $\mathbf{U}_k$, $k \geqslant i, j$, such that both U_i and U_j are subsets of U_k and $\preccurlyeq_i, \preccurlyeq_j, \lesssim_i, \lesssim_j$ are restrictions of $\preccurlyeq_k, \lesssim_k$ to U_i and U_j, respectively. The universal structure can be defined to be the union of all elements of the sequence, thus as a certain limit concept. One may view $\mathbf{U}_k$ as the concatenation of the structures $\mathbf{U}_i$ and $\mathbf{U}_j$, i.e. a structure such that $U_k = U_i \cup U_j$ and the set of the techniques and instruments applied in deciding hypotheses concerning the relations $\preccurlyeq_i, \preccurlyeq_j, \lesssim_i, \lesssim_j$ compose those relevant for $\preccurlyeq_k$ and $\lesssim_k$, respectively.

It is obvious that for such a concatenation to exist it is necessary that $\preccurlyeq_i, \preccurlyeq_j$ and $\lesssim_i, \lesssim_j$ coincide on $U_i \cup U_j$. The matter still requires further examination, but we must postpone it till the next section.

Observe that some measurement structures we are discussing

now (call them *M-structures*) may happen to be isomorphic with *CN*-structures discussed in the previous example. Imagine for instance a set A which consists of the objects $a_1, \ldots, a_{10}$ of equal mass and all their concatenations. Denote the object which results from two objects a, b by combining them together into a new one as $a \circ b$. The physical nature of this operation will be of minor interest for us, but we shall assume that the objects concatenated cannot possess a common part, i.e. concatenations such as $a_1 \circ a_1$, or $(a_1 \circ a_2) \circ (a_2 \circ a_3)$ do not exist. We shall also assume that the operation $\circ$ is symmetric and associative, i.e. $(a \circ b) \circ c = a \circ (b \circ c)$.

It may be easily verified that under the made assumptions the set A is equinumerous with the set $2^X - \{\emptyset\}$ considered in Example 1, and one has only to define $\preccurlyeq$, $\precapprox$ in an appropriate manner to have the measurement structure $\mathbf{A} = (A, \preccurlyeq, \precapprox)$ isomorphic with the structure $(2^X - \{\emptyset\}, \preccurlyeq, \precapprox)$, the latter being incidentally a measurement structure as well. The reader will easily reconstruct the formal conditions which have to be imposed on $\preccurlyeq$, $\precapprox$ and the concatenation operation in order for the structure $(A, \preccurlyeq, \precapprox)$ to have the desired properties.

The following theorem is easily seen to be valid.

THEOREM 4.2.1. *If*

(i) $(A, \preccurlyeq_A, \precapprox_A)$ *is a measurement structure,*

(ii) $(B, \preccurlyeq_B)$ *is a weakly ordered set,*

(iii) *the structure* $(A, \preccurlyeq_A)$ *is a substructure of* $(B, \preccurlyeq_B)$ (*i.e.* $A \subseteq B$) *and the relation* $\preccurlyeq_A$ *is the restriction of* $\preccurlyeq_B$ *to the elements of* A), and

(iv) *for each* b *in* B *there exists* a *in* A *such that* $a \sim b$ (*more exactly* $a \sim_B b$),

then there is exactly one relation $\preccurlyeq_B$ *in* $B \times B$ *such that* $(B, \preccurlyeq_B, \precapprox_B)$ *is a measurement structure.*

We shall precede the proof of Theorem 4.2.1. by proving the following

LEMMA 4.2.2. *Let* $\preccurlyeq$ *be a binary relation on* A *and* $\precapprox$ *a binary relation on* $A \times A$. *The structure* $(A, \preccurlyeq, \precsim)$ *is a measurement structure if and only if*

(i) *the relation* $\sim$ *defined to hold between* $a, b \in A$ *whenever* $a \preccurlyeq b$ *and* $b \preccurlyeq a$ *is a congruence in* $(A, \preccurlyeq, \precsim)$, *i.e. for all* a, a', b, b', c, c', d, d' *in* A,

(C0) $\sim$ *is an equivalence on* A,

(C1) *If* $a \sim a'$, $b \sim b'$, *then* $a \preccurlyeq b$ *if and only if* $a' \preccurlyeq b'$,

(C2) *If* $a \sim a'$, $b \sim b'$, $c \sim c'$, $d \sim d'$, *then* $(a, b) \precsim (c, d)$ *if and only if* $(a', b') \precsim (c', d')$,

and

(ii) *the quotient structure* $(A, \preccurlyeq, \precsim)/\sim$, *i.e. the structure* $(A^*, \preccurlyeq^*, \precsim^*)$ *defined as follows*:

a. $A^* = A/\sim\; = \{|a| : a \in A\}$, *where* $|a| = \{b \in A : a \sim b\}$,

b. $|a| \preccurlyeq^* |b|$ *if and only if* $a \preccurlyeq b$,

c. $(|a|, |b|) \precsim^* (|c|, |d|)$ *if and only if* $(a, b) \precsim (c, d)$, *is a measurement structure.*

PROOF. Assume first that $(A, \preccurlyeq, \precsim)$ is a measurement structure and let f be a homomorphism from $(A, \preccurlyeq, \precsim)$ into $(Re, \leqslant, L)$. Then we have $a \sim b$ if and only if $f(a) = f(b)$ and hence, by general properties of homomorphisms, $\sim$ must be a congruence in $(A, \preccurlyeq, \precsim)$. The mapping $f^* : A^* \to Re$ defined by the condition $f^*(|a|) = f(a)$ is a homomorphism from $(A^*, \preccurlyeq^*, \precsim^*)$ into **Re** and moreover $f(A) = f^*(A^*)$. By Corollary 4.1.11 and in view of the identity $f(A) = f^*(A^*)$, $(A^*, \preccurlyeq^*, \precsim^*) = (A, \preccurlyeq, \precsim)/\sim$ is a measurement structure if and only if $(f(A), \leqslant, L)$ is a measurement structure which in turn holds true if and only if $(A, \preccurlyeq, L)$ is a measurement structure. Since the last requirement is satisfied, $(A^*, \preccurlyeq^*, \precsim^*)$ is a measurement structure.

The presented argument can be reversed. If $\sim$ is a congruence in $(A, \preccurlyeq, \precsim)$, then the condition $f(a) = f^*(|a|)$ may serve as a definition of f, and if f^* is a homomorphism from $(A^*, \preccurlyeq^*, \precsim^*)$ into **Re**, then by the definition of f and conditions (C1), (C2) f is easily seen to be a homomorphism from $(A, \preccurlyeq, \precsim)$ into **Re**. Since again $f^*(A^*) = f(A)$ and $(f^*(A), \leqslant, L)$ is a measurement structure, so is $(A, \preccurlyeq, \underset{\sim}{\sqsubset})$. In this last concluding part of the argument again we have made use of Corollary 4.1.11.

Let us now turn back to Theorem 4.2.1 and present the proof of it in an outline.

Define $(a, b) \sqsubset_B (c, d)$ to hold true, whenever $(a', b') \underset{\sim}{\sqsubset}_A (c', d')$ for some a', b', c', d' in A such that $a \sim a', b \sim b', c \sim c', d \sim d'$. Conditions (iii) and (iv) and the definition of $\underset{\sim}{\sqsubset}_B$ allow us to prove that $\sim_B$ is a congruence in the structure $(B, \preccurlyeq_B, \underset{\sim}{\sqsubset}_B)$, and moreover the quotient structure $(B, \preccurlyeq_B, \underset{\sim}{\sqsubset}_B)/\sim_B$ coincides with the quotient structure $(A, \preccurlyeq_A, \underset{\sim}{\sqsubset}_A)/\sim_A$. Since the latter is a measurement structure in view of the assumption made and Lemma 4.2.2, thus making use of this lemma once again we arrive at the conclusion that $(B, \preccurlyeq_B, \underset{\sim}{\sqsubset}_B)$ is also a measurement structure.

Suppose now that for some relation $\underset{\sim}{\sqsubset}'_B$, $(B, \preccurlyeq_B, \underset{\sim}{\sqsubset}'_B)$ is a measurement structure. Then, in view of Lemma 4.2.2, $\sim_B$ is a congruence of that structure and thus for all a', b', c', d' such that $a \sim a'$, $b \sim b'$, $c \sim c'$, $d \sim d'$, $(a, b) \underset{\sim}{\sqsubset}'_B (c, d)$ if and only if $(a', b') \underset{\sim}{\sqsubset}'_B (c', d')$. In particular then the equivalence stated is valid for a', b', c', d' selected from A. Consequently, $(a, b) \underset{\sim}{\sqsubset}'_B (c, d)$ if and only if $(a, b) \underset{\sim}{\sqsubset} (c, d)$, which concludes the proof.

The importance of the theorem proved, which as a matter of fact is fairly trivial from the mathematical point of view, consists in displaying that the definition of a measurement structure does not presuppose any hypotheses which from the empirical point of view are dubious, regardless of whether the measurement structure is assumed to be finite or not. Thus, for instance, the assumption that an M-structure consists of infinite number of physical objects does not provide us with a sufficient base for proving that there are objects of arbitrary small or arbitrary great mass, or for proving that the values the parameter mass takes must form a dense or perhaps even continuous set. We may for instance imagine (which in view of Theorem 4.1.1 is perfectly possible) that the mass of any physical object is a multiple of a certain fixed object although we may as well imagine that the set of all actual measures $m(a)$ of the variable "mass" is dense or even coincides with *Re*. I am inclined to regard the relative independence of the definition of empirical assumptions as its advantage.

EXAMPLE 3. The last example I would like to present should allow us to grasp some other aspects of the measurements procedures.

Let me quote here Suppes' definition of a finite approximate measurement structure for beliefs (cf. Suppes, 1974) in its unchanged version.

DEFINITION 4.2.3. *A structure* $\mathbf{X} = (X, \mathscr{F}, \mathscr{S}, \geqslant)$ *is a* finite approximate measurement structure for beliefs *if and only if* X *is a non-empty set,* $\mathscr{F}$ *and* $\mathscr{S}$ *are algebras of sets on* X, *and the following axioms are satisfied for every* A, B *and* C *in* $\mathscr{F}$ *and every* S *and* T *in* $\mathscr{S}$:

Axiom 1. *The relation* $\geqslant$ *is a weak ordering of* $\mathscr{F}$;

Axiom 2. *If* $A \cap C = \varnothing$ *and* $B \cap C = \varnothing$, *then* $A \geqslant B$ *if and only if* $A \cup C \geqslant B \cup C$;

Axiom 3. $A \geqslant \varnothing$;

Axiom 4. $X > \varnothing$;

Axiom 5. $\mathscr{S}$ *is a finite subset of* $\mathscr{F}$;

Axiom 6. *If* $S \neq \varnothing$, *then* $S \geqslant \varnothing$;

Axiom 7. *If* $S \geqslant T$, *then there is a* V *in* $\mathscr{S}$ *such that* $S \sim T \cup V$.

Let me quote also Suppes' comments on the presented set of axioms. He writes:

> "In comparing Axioms 3 and 6, note that A is an arbitrary element of the general algebra $\mathscr{F}$, but event S (referred to in Axiom 6) is an arbitrary element of the subalgebra $\mathscr{S}$. Also in Axiom 7, S and T are standard events in the subalgebra $\mathscr{S}$, not arbitrary events in the general algebra. Axioms 1–4 are just the familiar de Finetti axioms without any change. Because all the standard events (finite in number) are also events (Axiom 5), Axioms 1–4 hold for standard events as well as arbitrary events. Axiom 6 guarantees that every minimal element of the subalgebra $\mathscr{S}$ has positive qualitative probability. Technically a *minimal element of* $\mathscr{S}$ is any event A in $\mathscr{S}$ such that $A \neq \varnothing$, and it is not the case that there is a non-empty B in $\mathscr{S}$ such that B is a proper subset of A. A minimal open interval (S, S') of $\mathscr{S}$ is such that $S < S'$ and $S' - S$ is equivalent to a minimal element of $\mathscr{S}$. Axiom 7 is the main structural axiom which holds only for the subalgebra and not for the general algebra; it formulates an extremely simple solvability condition for standard events".

Finite approximate measurement structures in Suppes' sense are not measurement structures in the sense accepted here. Nevertheless the affinity of the two concepts is obvious. To discuss the matter in more detail, given a finite approximate measurement structure $(X, \mathscr{F}, \mathscr{S}, \geqslant)$ consider its reduct $(\mathscr{S}, \geqslant)$. Since by Axiom 1, $\geqslant$ is a weak ordering of $\mathscr{F}$, $\geqslant$ is a weak ordering of the subset $\mathscr{S}$ of $\mathscr{F}$. Since $\mathscr{S}$ is a finite algebra of sets, the equivalence classes determined by its elements form a finite sequence

(1) $$|A_n| > |A_{n-1}| > \dots > |A_2| > |A_1|.$$

(We write $|A_i| > |A_j|$, when $A_i > A_j$). Note now that the interval (A_i, A_{i-1}) must be minimal. Consider for instance the interval (A_2, A_1). Since $A_2 > A_1$, then by Axiom 7 there exists an element C such that $A_2 \sim A_1 \cup C$. Clearly this element must belong to the class $|A_1|$. In turn, by the same argument we conclude that $A_3 \sim A_2 \cup C$. Suposse that C is in $|A_2|$. Then it can be presented in the form $C_1 \cup C_2$, where $C_1 \cup C_2$ are in A_1 and $C_1 \cap C_2 = \varnothing$. But, were it possible, there would exist an element $A_2 \cup C_1$ such that $A_3 > A_2 \cup C_1 > A_2$, which contradicts our assumption that A_3 is the next element greater than A_3.

Put $d(A_i, A_j) = |i-j|$, and define $(A_i, A_j) \precsim (A_k, A_j)$ if and only if $d(A_i, A_j) \leqslant d(A_k, A_l)$. The definition in an obvious way can be extended onto all elements of $\mathscr{S}$ not only those which were used for defining the elements of the sequence (1). Namely, we define $(A, B) \precsim (C, D)$ if and only if $(A_i, A_j) \precsim (A_k, A_l)$, where $A_i \sim A$, $A_j \sim B$, $A_k \sim C$, and $A_l \sim D$.

ASSERTION 4.2.4. *If $(X, \mathscr{F}, \mathscr{S}, \geqslant)$ is a finite approximate measurement structure for beliefs, then $(\mathscr{S}, \geqslant, \precsim)$, where $\precsim$ is the relation on $\mathscr{S} \times \mathscr{S}$ defined in the way described above, is a measurement structure.*

The proof of this assertion is left to the reader who may also omit it. It is not difficult but involves many tedious technicalities of no significance for further considerations.

Observe that the measurement structure $(\mathscr{S}, \geqslant, \precsim)$ is of "the same kind" as the structures discussed in Examples 1 and 2. The difference does not concern the techniques of constructing the probability function P, i.e. the homomorphism from $(\mathcal{S}, \geqslant, \precsim)$ into $(Re, \geqslant, L)$ (note that the direction of the first ordering relation has been reversed), but the experiments by means of which hypotheses about the order of elements of $\mathcal{S}$ can be decided. To put it in other words, when the operational definitions of the ordering relations involved in the discussed structures are given, the mathematical problems left to be solved in order to define n, m or P (depending on the structure in question) are exactly the same.

The following theorem, being a variant of a theorem stated and proved by P. Suppes (1974, p. 169) throws some light on the role of measurement structures "hidden" in finite approximate measurement structures for beliefs.

ASSERTION 4.2.5. *Let* $\mathbf{X} = (X, \mathcal{F}, \mathcal{S}, \geqslant)$ *be a finite approximate measurement structure for beliefs. Then there exists a probability measure* P *on* $\mathcal{F}$ *such that for any two events* S, T

$$S \geqslant T \text{ if and only if } P(S) \geqslant P(T)$$

and if P_1, P_2 *are such probabilities, then for all* S *in* $\mathcal{F}$

(i) $|P_1(S) - P_2(S)| \leqslant \frac{1}{n}$, *where* n *is the number of minimal elements in* $\mathcal{S}$, *and*

(ii) $P_1(S) = P_2(S)$, *for all* S *in* $\mathcal{S}$.

The reader can consult P. Suppes (1974, p. 170) for the details which may be useful in proving the assertion. But the main idea is simple. When $S \in \mathcal{F}$, one should localize S by comparing it with the elements $A_1, \ldots, A_n$ involed in (1) or any elements equivalent to them. If $A_i > S > A_{i-1}$, all we can do is to put

$$P(S) = \frac{P(A_i) + P(A_{i-1})}{2} \pm \frac{1}{2n}.$$

The measurement system with which we operate does not allow us to determine P in a more precise way. A further discussion of approximate measurement will be carried out in the next section.

4.3. APPROXIMATE MEASUREMENT

To begin with let us define the notion of an approximate measurement in a fully general way.

DEFINITION 4.3.1.

A. *A structure* $\mathbf{A} = (A, U, \preccurlyeq, \sqsubsetapprox)$ *is an* approximate measurement structure *if and only if the following conditions are satisfied*:

(AM1) *U is a non-empty subset of A.*

(AM2) $\preccurlyeq$ *is a weak ordering relation defined on A.*

(AM3) $(U, \preccurlyeq, \sqsubsetapprox)$ *is an interval measurement structure.*

B. *If moreover U is a finite set, the structure* **A** *will be said to be a* finite approximate measurement structure.

Assertion 4.2.4. brings to light how the notion of an approximate measurement structure is related to that of a finite approximate measurement structure for beliefs. Although in an implicit way, the latter concept serves as an exemplification of the concept defined now and, at the same time, as a pattern for it. In particular, Assertion 4.2.4. can be generalized as follows.

THEOREM 4.3.2. *If* $(A, U, \preccurlyeq, \sqsubsetapprox)$ *is a finite approximate measurement structure, then*:

(i) *there exists a function* $f\colon A \to Re$ *such that for all* a, b in A,

$$a \preccurlyeq b \quad \textit{if and only if} \quad f(a) \leqslant f(b),$$

(ii) *If* f_1, f_2 *are such functions, and for some* e_0, e_1, e_2 *in* U, *such that not* $e_1 \sim e_2$,

$$f_1(e_0) = f_2(e_0),$$

and

$$|f_1(e_1) - f_1(e_2)| = |f_2(e_1) - f_2(e_2)|,$$

then for each a in A such that for some b, c *in* $U, b \preccurlyeq a \preccurlyeq c$,

$$|f_1(a) - f_2(a)| \leqslant \frac{1}{k},$$

where $k+1$ *is the greatest number of elements in U by means of which a chain of the form* $e_1 \prec e' \prec e'' \prec \ldots \prec e_2$ *can be formed. Moreover,*

(iii) *If* $a \in U$, *then* $f_1(a) = f_2(a)$.

PROOF. The set U is finite and therefore equivalence classes determined by its elements form a finite sequence

(1) $$|a_0| \prec |a_1| \prec \ldots \prec |a_n|.$$

As usual we write $|a_i| \prec |a_j|$ when $a_i \prec a_j$. The most natural scale f for measuring the quantity defined by the measurement structure $(U, \preccurlyeq, \precsim)$ seems to be the one whose both the origin and unit are selected so as to have $f(a_0) = 0$ and $|f(a_0) - f(a_1)| = 1$. Clearly some practical considerations may suggest another decision, but in order to simplify the proof carried out now we shall select f in the way described.

Note that under the choice made, we have

(2) $$f(a_k) = k,$$

for all $k = 0, 1, \ldots, n$. Indeed we have $f(a_0) = 0$ by the assumption, and suppose now that $f(a_{k-1}) = k-1 \geqslant 1$. $(U, \preccurlyeq, \precsim)$ is a measurement structure and since $(a_0, a_1) \precsim (a_0, a_{k-1}) \precsim (a_0, a_k)$, there is a standard sequence

$$a_0, a', \ldots, a'_{k-1}, a'_k, \ldots, a'_n$$

such that $(a_i, a_{i+1}) \approx (a_0, a_1)$, $(a_0, a_{k-1}) \approx (a_0, a_k)$ and $(a_0, a'_{n-1}) \precsim (a_0, a_k) \precsim (a_0, a'_n)$. Suppose that $n \neq k$, i.e. $n > k$, then $(a_0, a'_k) \sqsubset (a_0, a_k) \precsim (a_0, a'_n)$ and therefore $a'_k \prec a_k$. At the same time however we clearly have $a_{k-1} \prec a'_k$ and, contrary to the assumption made, it turns out that there is an element a'_k in between a_{k-1} and a_k. The argument has been presented in an outline, but the reader should have no difficulty with completing it. The remaining part of the proof will also have an abbreviated form.

Select now any a in A such that for some b, c in U, $b \prec a \prec c$. If $|f(b) - f(c)| > 1$, then in view of what we have established above there are $b_1 \prec b_2 \prec \ldots \prec b_k$ such that $b = b_1$, $b_k = c$ and $|f(b_i) - f(b_{i+1})| = 1$. Suppose that $b_k \preccurlyeq a \preccurlyeq b_{k+1}$. Clearly any extension of f' onto A for which the requirement

$$a \preccurlyeq b \quad \text{if and only if} \quad f''(a) \leqslant f''(b)$$

is satisfied must be such that

$$f'(b_k) \leqslant f'(a) \leqslant f'(b_{k+1}),$$

i.e.

$$k \leqslant f'(a) \leqslant k+1.$$

Let f_*, f^* be two extensions of f of utterly arbitrary values. Let us put $f_*(a) = k$ and $f^*(a) = k+1$. We have then

$$|f_*(a) - f^*(a)| = 1.$$

Thus in general, given any two extensions f_1, f_2 of f onto A, we must have

(3) $$|f_1(a) - f_2(a)| \leqslant 1.$$

Observe now that since both e_1 and e_2 are in U, then $f_1(e_1) = f_2(e_1) = f(e_1)$ and similarly $f_1(e_2) = f_2(e_2) = f(e_2)$. Suppose that

(4) $$|f(e_1) - f(e_2)| = k.$$

Clearly, this means that for some a_i, a_{i+k}, $e_1 \in |a_i|$ and $e_2 \in |a_{i+k}|$, and thus $k+1$ is the greatest number of elements in U which can be exploited in order to form a chain $e_1 \prec e' \prec e'' \prec \ldots \prec e_2$ described in the theorem.

Transform the scale f_1, f_2 to f_1', f_2' so that

$$f_1'(e_0) = f_2'(e_0) = \xi_0,$$

and

$$|f_1'(e_1) - f_1'(e_2)| = |f_2'(e_1) - f_2'(e_2)| = \delta.$$

In view of (4), in order to obtain such scales we must put

$$f_1'(a) = \frac{1}{k}\delta f_1(a) - l,$$

and

$$f_2'(a) = \frac{1}{k}\delta f_2(a) - l,$$

where $l = |f(a_0) - f(e_0)|$.

Observe now that

$$|f_1'(a) - f_2'(a)| = \frac{1}{k}\delta . |f_1(a) - f_2(a)|$$

which, by (3), gives

$$|f_1'(a)-f_2'(a)| \leqslant \frac{1}{k}\delta,$$

concluding the proof.

Although the proved theorem makes it clear in what sense approximate measurement structures may be treated as mathematical models for approximate measurements, still the adequacy of these models can be questioned.

What affects accuracy of the measurement is, in the first place, fuzziness of the ordering relations, $\preccurlyeq$, $\precsim$ defined in an operational manner. If differences between objects or in intervals are too small, the operational procedures by means of which $\preccurlyeq$ and $\precsim$ are defined may happen not to allow us to decide whether $a \preccurlyeq b$ or $b \preccurlyeq a$ or perhaps $a \sim b$, and similarly whether $(a, b) \precsim (c, d)$, $(c, d) \precsim (a, b)$ or $(a, b) \approx (c, d)$.

The facts we mentioned are reflected to some extent by the definition of an approximate measurement structure. Namely, given such a structure $(A, U, \preccurlyeq, \precsim)$, one may extend the definition of $\precsim$ onto at least some of the intervals in $A \times A$. Note for instance that if a, b are in U, a', c are in A and $a \sim a'$, $a \preccurlyeq b \prec c$, then this clearly justifies postulating that $(a, b) \sqsubset (c, d)$.

The principle on which the procedure of expanding $\precsim$ onto $A \times A$ is based is transparent. We are looking for such a definition of $\precsim$ under which $(A, \preccurlyeq, \precsim)$ is an interval structure and the restriction of $\precsim$ to $U \times U$ coincides with $\precsim$ defined in the original way. It may well happen however that we shall not be able to decide about some intervals how they are related among themselves. For instance intervals shorter than the minimal intervals in $U \times U$ may turn out to be uncomparable.

The indefiniteness of $\precsim$ expanded onto $A \times A$ yields the same results as fuzziness of $\precsim$ defined in an operational way, in particular affects the preciseness of the measurement. At the same time, however, the "sharpness" of $\preccurlyeq$ produces a paradoxical effect. Although in general the results of measurements need not be accurate, still a precise measure can be assigned to each object

a such that $a \sim b$, for some b in U. This can hardly be considered to be characteristic of any approximate measurement!

In order to arrive at a more realistic model for an approximate measurement we must somehow define measurement structures so as to make them involve ordering relations modelling the fuzziness of $\preccurlyeq$ and $\underset{\sim}{\sqsubseteq}$.

DEFINITION 4.3.3. *A structure* $\mathbf{A} = (A, U, \dot{\prec}, \dot{\sim}, \dot{\sqsubset}, \dot{\approx})$ *is an* operational measurement structure *if and only if the following conditions are satisfied:*

OM1. *U is a finite subset of A consisting of at least two elements.*

OM2. *$\dot{\prec}$ is a binary transitive and asymmetric relation on A.*

OM3. *$\dot{\sim}$ is a binary, reflexive, and symmetric relation on A.*

OM4. *$\dot{\sqsubset}$ is a binary, transitive, and asymmetric relation on $A \times A$.*

OM5. *$\dot{\approx}$ is a binary, reflexive, and symmetric relation on $A \times A$.*

OM6. *For all a, a', b in A:*

a. *$a \dot{\prec} b$ or $b \dot{\prec} a$ or $a \dot{\sim} b$,*

b. *If $a \dot{\sim} a', b \dot{\sim} b'$ and $a \dot{\prec} b$, then $a' \dot{\preccurlyeq} b'$, i.e. either $a' \dot{\prec} b'$ or $a' \dot{\sim} b'$.*

OM7. *The structure $(U, \dot{\preccurlyeq}, \dot{\underset{\sim}{\sqsubseteq}})$, where $\dot{\preccurlyeq}$ is defined as in* AM6b, *and $\dot{\underset{\sim}{\sqsubseteq}}$ is to be defined in a similar way, is an interval measurement structure.*

OM8. *If $|a_0| \dot{\prec} |a_1| \dot{\prec} \dots \dot{\prec} |a_n|$ is the chain of all equivalence classes in $U/\dot{\sim}$ (we write $|a_i| \dot{\prec} |a_j|$ whenever $a_i \dot{\prec} a_j$), then for all a, a_i', a_j' in A,*

a. *If $a \dot{\sim} a_i$, then $a_{i-1} \dot{\prec} a$, $i = 1, \dots, n$,*

b. *If $a \dot{\sim} a_i$, then $a \dot{\prec} a_{i+1}$, $i = 0, \dots, n-1$,*

c. *If $a_i \dot{\sim} a_i', a_j \sim a_j'$ $(i, j = 0, 1, \dots, n)$, then $a_i \dot{\prec} a \dot{\prec} a_j$ if and only if $a_i' \dot{\prec} a \dot{\prec} a_j'$.*

OM9. *There exists a measurement structure $(A, \preccurlyeq, \underset{\sim}{\sqsubseteq})$ such that for all a, b, c, d in A,*

a. *If $a \dot{\prec} b$, then $a \prec b$,*

b. *If $a \sim b$, then $a \dot{\sim} b$,*

c. *If* $(a, b) \dot{\sqsubset} (c, d)$, *then* $(a, b) \sqsubset (c, d)$,

d. *If* $(a, b) \approx (c, d)$, *then* $(a, b) \dot{\approx} (c, d)$,

e. *There exists a chain*

$$a_0^* \prec a_1^* \prec \ldots \prec a_n^*$$

of elements of A *which corresponds to the chain* $|a_0| \dot{\prec} |a_1| \dot{\prec} \ldots \dot{\prec} |a_n|$ *of all equivalence classes in* $U/\sim$ *such that* $a_i \dot{\sim} a_i^*$ *for all* $i = 0, 1, \ldots, n$ *and* $(a_i^*, a_j^*) \approx (a_k^*, a_l^*)$ *if and only if* $(a_i, a_j) \dot{\approx} (a_k, a_j)$.

THEOREM 4.3.4. *If* $(A, U, \dot{\prec}, \dot{\sim}, \dot{\sqsubset}, \dot{\approx})$ *is an operational measurement structure, and*

$$|a_0| \dot{\prec} |a_1| \dot{\prec} \ldots \dot{\prec} |a_n|$$

is the chain of all equivalence classes in $U/\dot{\sim}$ *then there exists a function* $f\colon A \to Re$ *such that the following conditions hold true*

(i) *for each* $k = 0, 1, \ldots, n$ *there exists* $a'_k \dot{\sim} a_k$ *such that* $f(a'_k) = k$.

(ii) *If* $a \dot{\prec} b$, *then* $f(a) < f(b)$.

(iii) *If* $(a, b) \dot{\sqsubset} (c, d)$, *then* $|f(a)-f(b)| < |f(c)-f(d)|$.

Moreover, if f_1, f_2 *are two such functions, then for each* a *in* A *such that for some* b, c *in* U, $b \dot{\prec} a \dot{\prec} c$,

$$|f_1(a)-f_2(a)| < 2.$$

PROOF (*an outline*). Let $(A, \preccurlyeq, \underset{\approx}{\sqsubset})$ be the measurement structure whose existence is guaranteed by OM9 and let f be the homomorphism from $(A, \preccurlyeq, \underset{\approx}{\sqsubset})$ into **Re** defined by the conditions: $f(a_0^*) = 0, f(a_1^*) = 1$, where a_0^*, a_1^* are elements of the chain whose existence is assured by OM9e. The conditions imposed on both that chain and the function f imply that

(1) $$f(a_k^*) = k,$$

and since $a_k^* \dot{\sim} a_k$, f satisfies the requirement (i). Clearly the requirement (ii) is also satisfied since $a \dot{\prec} b$ implies $a \prec b$ in virtue of OM9a. The same argument but referring to OM9c establishes (iii).

Consider now any function g which satisfies conditions being counterparts of those imposed on f. Note that g can be defined with respect to a chain $a_0^+ \prec a_1^+ \prec \ldots \prec a_n^+$ different from the "star" chain which has been employed to construct f. Obviously the "cross" chain must satisfy the analogon of OM9e, which incidentally entails neither that $a_i^* \sim a_i^+$ nor even that $a_i^+ \mathrel{\dot{\sim}} a_i$. We only have $a_i^* \mathrel{\dot{\sim}} a_i$ and $a_i^+ \mathrel{\dot{\sim}} a_i$. We also have

(2) $\qquad g(a_k^+) = k.$

Let $a \in A$ and let for some b, c in U, $b \mathrel{\dot{\prec}} a \mathrel{\dot{\prec}} c$. In order to conclude the proof we have to show that

(3) $\qquad |f(a) - g(a)| \leqslant 2.$

Let a_{m-1} be the greatest element among all a_i's, $i = 0, 1, \ldots, n$, such that $a_{n-1} \mathrel{\dot{\prec}} a$. Then by OM6a, either $a \mathrel{\dot{\sim}} a_m$ or $a \mathrel{\dot{\prec}} a_m$. Suppose first that $a \mathrel{\dot{\sim}} a_m$. We have then

$$a_{m-1} \mathrel{\dot{\prec}} a \mathrel{\dot{\prec}} a_{m+1},$$

and thus by AM8c

$$a_{m-1}^* \mathrel{\dot{\prec}} a \mathrel{\dot{\prec}} a_{m+1}^*$$

and similarly

$$a_{m-1}^+ \mathrel{\dot{\prec}} a \mathrel{\dot{\prec}} a_{m+1}^+.$$

The inequalities stated above entail (by OM9a):

$$a_{m-1}^* \prec a \prec a_{m+1}^*$$

and

$$a_{m-1}^+ \prec a \prec a_{m+1}$$

By (1) and (2) we obtain then

$$m-1 < f(a) < m+1$$

and

$$m-1 < g(a) < m+1$$

which directly yields (3).

If $a_{m-1} \mathrel{\dot{\prec}} a \mathrel{\dot{\prec}} a_m$, the argument runs in exactly the same way and allows eventually to establish that

$$m-1 < f(a) < m,$$

and

$$m-1 < g(a) < m,$$

which gives even better approximation than the previous one.

Instead of commenting on the notion of an operational measurement structure let us define the notion of an operational scale and state a few more theorems.

DEFINITION 4.3.5. *Let* **A** $= (A, U, \precsim, \sim, \dot{\sqsubset}, \approx)$ *be an operational measurement structure.*

A. *A function* $\Phi: A \to [Re]$, *where* $[Re]$ *is the set of all non-empty closed intervals of real numbers will be said to be an* operational scale for **A** *if and only if for some* $\xi_0, \delta \in Re, \delta > 0$ *the following condition is satisfied. For each* a *in* A *and for each* x *in* Re, $x \in \Phi(a)$ *if and only if:* (1) *for each* $\varepsilon > 0$ *there exists* x_0 *in* Re *such that* $|x - x_0| < \varepsilon$, *and* (2) *for some function* f *which satisfies the conditions* (i)-(iii) *stated in* Theorem 4.3.4

$$\frac{1}{\delta} f(a) + \xi_0 = x_0 .$$

B. *The numbers* ξ_0, δ *will be called the* origin *and the* unit *respectively* of the operational measure Φ.

The proofs of the theorems we are now going to state will be omitted.

THEOREM 4.3.6. *If* **A** *is an operational measurement structure and* Φ_1, Φ_2 *are operational scales for* **A** *having the same origins and the same units, then* $\Phi_1 = \Phi_2$.

THEOREM 4.3.7. *Let* **A** *be an operational measurement structure,* Φ *an operational scale for* **A**. *Then for all* $\alpha, \beta \in Re, \beta > 0$, *the function* $\Phi: A \to [Re]$ *(A being the set of elements of* **A**) *defined by the condition: for each* a *in* A *and each* x *in* Re,

$$x \in \Phi(a) \quad \textit{iff} \quad \frac{x}{\beta} + \alpha \in \Phi(a),$$

is an operational scale for **A**.

COROLLARY 4.3.8. *If* **A** *is an operational measurement structure, then for each* ξ_0 *and each* δ *in* Re, $\delta > 0$ *there is a scale* Φ *of* A *such that* ξ_0 *is the origin and* δ *is the unit of* Φ.

THEOREM 4.3.9. *Let* **A** $= (A, U, \precsim, \sim, \dot{\sqsubset}, \approx)$ *be an operational measurement structure and let for some superset* U' *of* U, **A**$' =$

= $(A, U', \dot{\prec}, \dot{\sim}, \dot{\sqsubset}, \dot{\approx})$ *be also an operational measurement structure. If* Φ *is an operational scale for* **A**, Φ' *is an operational scale for* **A**$'$ *and* Φ, Φ' *have both a common origin and unit, then for each* a *in* A

$$\Phi'(a) \subseteq \Phi(a).$$

(*If for some* a *in* A, $\Phi'(a) \subset \Phi(a)$, *the structure* **A**$'$ *will be said to be* finer *than* **A**).

THEOREM 4.3.10. *If structures* $\mathbf{A}_i = (\mathbf{A}, U_i, \dot{\prec}, \dot{\sim}, \dot{\sqsubset}, \dot{\approx})$, $i = 1, 2, \ldots$, *form an infinite sequence of operational measurement structures such that* $U_i \subset U_j$ *for* $i < j$, *and moreover for any two* a, b *in* A *there exists a structure* $\mathbf{A}_i$ *in the sequence such that for some* e_1, e_2, e_3 *in* U_i,

$$e_1 \dot{\prec} a \dot{\prec} e_2 \dot{\prec} b \dot{\prec} e_3,$$

then there exists exactly one measurement structure $(A, \preccurlyeq, \sqsubseteq)$ *such that the condition* OM9 *of Definition* 4.3.3 *is satisfied for all structures* $\mathbf{A}_i$.

Theorem 4.3.10 makes it clear in what sense measurement structures can be treated as a limit case of operational measurement structures, and consequently also in what sense the scale for a measurement structure **A** can be treated as the limit case of an operational scale.

4.4. THEORETICAL VERSUS OPERATIONAL CONCEPTION OF MEASUREMENT

The measurement theory which has been presented in this chapter is of quite special nature. We have dealt neither with measurement techniques nor even with methodology of measurement. Still "theory of measurement" is a commonly accepted name for the considerations whose modest sample can be found in the previous sections of this chapter.

The objective of the theory of measurement are relations between quantitative and qualitative concepts. Its main task is to bring out how the quantitative concepts can be reduced to qualitative ones, or in other words, how qualitative relations characteristic of empirical phenomena can be presented and examined in the form of quantitative ones. The objective of this theory is then the intrinsic nature of measurement rather than any problems of practical origin connected with implementation of measurement procedures.

This specific nature of the theory of measurement is responsible for the fact that the very conception of measurement characteristic of this theory differs considerably from the one the practician is likely to accept. For him the measurement is a sequence of operations performed in accordance with certain fixed (though not necessarily codified) rules. He is inclined to view measurement as practical activity rather than a specific sort of theorizing. Let us discern between those two conceptions of measurement labelling them theoretical and operational respectively. In what follows we shall try to examine the two conceptions closer though in a loose manner, and to find out how they are related to each other.

A suitable example should be of some help for our further discussion. Suppose that the objective of an experiment is to establish whether two momentary events e_1, e_2 are simultanous. Denote the time of event e_1 as t_1 and the time of e_2 as t_2. (t_1, t_2 and all the quantities dealt with in this example will be assumed to be measured with respect to a fixed frame of reference $\mathscr{K}$). Let s_1 be the position in which the event e_1 takes place, and s_2 the position of e_2. The hypothesis to be verified is

(h) $\quad t_1 = t_2,$

or more exactly

(h_ε) $\quad t_1 = t_2 \pm \varepsilon$

where ε is the presupposed degree of accuracy of the experiment.

The "traditional" method of measuring time would consist in making use of two clocks which after synchronizing them would be transported to s_1 and s_2 respectively. From the point of view of relativistic mechanics the procedure described is not acceptable (the transposed clocks would slown down) and we have to find another way. There are many possible solutions to the problem, let us discuss one of them.

The experimenter finds a point s_0 such that

(1) $$|s_0 - s_1| = |s_0 - s_2|,$$

i.e. the distance between s_0 and s_1 coincides with that between s_0 and s_2. Next, at time t_1 he sends a light signal from s_1 to s_0 and at time t_2 he sends a light signal from s_2 to s_0. Let t'_1 and t'_2 be the times at which the signals from s_1 and s_2, respectively, reach s_0. Hypothesis (h_ε) holds true if and only if

(h^*_ε) $$t'_1 = t'_{2 \pm \varepsilon}.$$

The procedure described is sometimes claimed to provide a classical example of an operational definition of simultaneity of distant events. I am afraid that the description of the procedure to be implemented in order to decide (h) is too schematic to deserve to be called an operational definition, at least if an operational definition is expected to characterize the sequence of operations characteristic of a given measurement procedure in a unique way (up to certain "inessential" details). Note for instance that there are many ways in which the experimenter may select the point s_0 so that (1) is valid, note also that we have not described the way in which the hypothesis (h^*_ε) is to be decided. If the procedure described defines simultaneity, it does so not because it amounts to an operational definition but rather because it is based on a law which defines simultaneity of distant events in terms of simultaneity of events localized at the same place. The law is as follows:

(2) *If* $|s_0 - s_1| = |s_0 - s_2|$, *then* ($h_\varepsilon$) *if and only if* ($h^*_\varepsilon$).

(It is worth mentioning perhaps that since (2) is a conditional, it should be treated as a meaning postulate rather than a definition in a proper sense of the word).

Neither the selection of s_0 nor deciding (h^*_ε) has to be a trivial problem. Let us discuss briefly the latter. Suppose first that the experimenter registers the times t'_1, t'_2 of arrivals of the signals without any instruments, he just observes the flashes of light on a screen located at s_0. If the interval $t'_1 - t'_2$ is relatively very small, the method is useless; the human eye would not be able to catch the difference between the time instants t'_1, t'_2 and both the flashes will be seen as simultaneous even if $|t'_1 - t'_2|$ is greater than the presupposed degree of accuracy ε.

An alternative method may consist in photographing the flashes on a film reel set in a very fast motion. The degree of accuracy available with this method is considerably higher than in the previous case, but if both $|t'_1 - t'_2|$ and ε are small, it may still turn out to be unsatisfactory, and a further improvement may be necessary. One may try, for instance, to exploit the phenomenon of interference of polarized light in a way similar to that employed in the Michelson experiment.

The example discussed allows to grasp in a sufficiently clear way the differences between theoretical and practical (operational) problems each experiment involves (but perhaps the simplest ones). The task of a theoretician, who certainly may be a practician at the same time, is to design the experiment by defining the schema of it. By a *schema of an experiment* we shall understand a mathematically provable conditional of the form

(*) *If* $\lambda_1, \ldots, \lambda_m, \gamma_1, \ldots, \gamma_n$, *then* $\alpha^* \leftrightarrow \alpha$,

where $\lambda_1, \ldots, \lambda_m$ are empirical laws, $\gamma_1, \ldots, \gamma_n$ are statements which describe the situations to be produced by the experimenter, α is a tested hypothesis and α^* is a test equivalent to α.

In general, formulas involved in a schema of an experiment depend on certain parameters $d_1, \ldots, d_n$ which, when fixed, define an application of the schema. For instance the parameters involved (implicitly) in the schema (2) are symbols of events e_1, e_2.

If α is an elementary formula, schema (*) will be referred to as a *schema of an elementary experiment*. If furthemore α is of the form $F(t, a_1, \ldots, a_k) = x$, we shall refer to (*) as the *schema of an ideal experiment*. In its ideal form the experiment we have described falls under the schema

(3) *If* (2) *and* (1), *then* (h*) $\leftrightarrow$ (h).

It is perhaps not immediately seen that (h) is of the form

$F(t, a_1, \ldots, a_k) = x$, but t_1, t_2 stand for $t(e_1)$ and $t(e_2)$, where t is the quantity "time".

Whenever an elementary experiment is not just planned but actually carried out, a tested hypothesis must take the form $F(t, a_1, \ldots, a_k) = x \pm \varepsilon$, where ε defines the degree of accuracy the experimenter wants to achieve. It is a matter of mathematical calculations to verify whether approximate data the experimenter is able to collect suffice to prove the tested hypothesis in its approximate form. In what follows an approximate counterpart of a tested hypothesis α will be occasionally denoted as α_ε, where ε is the accuracy desired.

Each application of an experimental schema leads unavoidably to the problem of deciding some empirical hypotheses different from the tested one. Thus for instance in order to decide (h) one has to decide (h*). The statement (1) which is also involved in the schema (3) is a hypothesis of special sort because this is just the experimenter's task to take care to select s_0 so that (1) holds true. Still when s_0 is selected, the verification problem arises, the experimenter is to check somehowwthat his selection suits the purpose. In that way the original experimental schema may give birth to some "subschemas" either of the form

$$\textit{If } \lambda_1', \ldots, \lambda_k', \gamma_1', \ldots, \gamma_l', \textit{ then } \alpha^{**} \leftrightarrow \alpha^*,$$

or of the form

$$\textit{If } \lambda_1'', \ldots, \lambda_s'', \gamma_1'', \ldots, \gamma_r'', \textit{ then } \delta_1 \leftrightarrow \gamma_1.$$

Clearly, the three schemas can be combined together into one of the form:

$$(**) \qquad \begin{aligned} &\textit{If } \lambda_1, \ldots, \lambda_m, \lambda_1', \ldots, \lambda_k', \lambda_1'', \ldots, \lambda_s'' \\ &\textit{and } \delta_1, \gamma_2, \ldots, \gamma_n, \gamma_1', \ldots, \gamma_l', \gamma_1'', \ldots, \gamma_r'', \\ &\textit{then } \alpha^{**} \leftrightarrow \alpha. \end{aligned}$$

Schema (**) will referred to as a *specification of* (*).

It is sometimes suggested that in order for a schema of an experiment to be put in an ultimate form, i.e. one which does not require further specifications, both the hypothesis which serves as a test equivalent to the tested hypothesis and the hypothesis which refers to a state of affairs to be produced by the experimenter should be couched in a commonly intelligible language (For instance R. Giles wants the language in which experimental procedures are described to be so simple that "it could be understood even by a child" cf. 1975 p. 9). Postulating that in its ultimate wording the schema of an experiment should involve apart from laws only observational sentences is still another variant of this demand.

The common intelligibility demand seems to me completely unrealistic. Many of the situations the scientist deals with are so considerably different from those known from everyday experience that they simply cannot be described without employing special terms whose meaning cannot be grasped without suitable training and thus they cannot be understood equally well by the layman and by the expert.

What seems to be a realistic demand is that experiment schemas should involve apart from laws only such formulas which are decidable by standarized well known techniques. Call the schemas which satisfy this demand *ultimate*. Usually there is not much sense in discussing whether a given experiment schema is ultimate or not unless both a particular application of the schema is considered and a degree of accuracy is presupossed. As we have already seen the same schema may be ultimate when the accuracy postulated is sufficiently low and require further specifications when a greater accuracy is demanded.

The possibility to reduce quantitative hypotheses to qualitative ones offered by the theory of measurement may suggest that in its ultimate formulation a schema of an experiment should involve quantitative terms at most in the tested hypothesis and in the empirical laws appearing in it. An examination of an experiment shows that all the hypotheses which the experimenter decides in the course of an experiment are either qualitative or may be viewed as conclusions drawn from such with the help of a suitable experiment schema. The significance of the theory of measurement lies just in the fact that it establishes links between the quantitative theories and qualitative experiments.

It would be very impractical, however, to demand experimental schemas to be worded in quantitative terms (except for their "theoretical" parts). There are at least two reasons for that.

In its fundamental form a schema of an experiment is usually a schema of an ideal experiment. When a schema for α is given, say just in the form (*), schemas for α_ε are derivable from the former by means of mathematical transformations. One must calculate how the inaccuracies in deciding the hypothesis $\gamma_1, \ldots, \gamma_n$, α^* affect the accuracy of measurement of α. The quantitative form of the schema facilitates calculations, if not makes them possible at all.

The second reason to be mentioned is as follows. Apart from being flexible under mathematical transformations, a good experimental schema should leave enough room for the experimenter's decision how to carry out the experiment in details. One certainly may reduce the hypotheses like "the distance between a and b equals 5.2 ± 0.2 m" or "the temperature of air at time t in s is 22 ± 1 C°" or "the mass of a equals $0{,}023 \pm 0{,}004$ mg" to their qualitative counter parts, and certainly it is desirable to know how precisely this can be done, but the experimenter is able to select in that practical situation he faces, the best way to decide them, and there are no reasons to limit his possibilities by unnecessary specifications of the experimental schema.

Thus far we have mainly been concerned with theoretical aspects of an experiment, let us in turn say a few words on operational ones. The operational problems an experiment involves concern either the technique of carring out the experiment or the criteria by means of which the experimenter controls at each stage of the experiment whether conditions necessary for the experiment to be valid are satisfied.

Note that already the very fact that the experiment is based on certain laws implies that the experimenter should check whether the circumstances in which the experiment is carried out allow making use of these law. Thus for instance, in the case of schema (3), in order for (2) (being the only law which (3) involves) to be true, velocities of the light signals sent from s_1 and s_2 should be the same. Clearly one may try to test the hypothesis concerning the velocity of light signals by means of a separate experiment, but the sequence of experiments may not be continued ad infinitum and some hypotheses must be accepted "a priori", i.e. without a conclusive evidence.

The a priori assumptions intervene practically in all experiments. Since our main example is quite sophisticated, consider the following one. In many experiments which involve animals it is of great significance to keep them away from anything which might cause their excitement or anxiety. Usually, however, the experimenter is able to conclude only indirectly, on the base of some random observations of the behaviour of animals and on the base of his belief in accuracy of his own instructions and carefulness of animal's guardian, that this condition is satisfied. The hypothesis "there have been no significant stressors affecting the behaviour of animals" might certainly be subjected to standarized empirical verification but the experimenter's possibilities are limited and he must to some extent rely on imagination, intuition, past experience (whatever else you might call the experimenter's ability to judge the situation he deals with in an adequate way) in dividing all hypotheses relevant for the experiment into those which are to be carefully verified and those which will be accepted "a priori" mainly on the base of his personal (which does not mean unmotivated) beliefs.

Clearly the problem is connected not only with the laws involved in an experiment, but also with its technical aspect. The experimenter is not able to verify all physical and/or chemical, parameters characteristic of instruments, reagents, specimens, laboratory equipment etc. he applies. "I believe John, the guardian of animals" and "I trust Mr Smith, the producer of reagents" can hardly stand for scientific justifications of the claim that the experiment is valid. Still the experimenter cannot help making use of such justifications.

What makes the experimenter trust himself and trust others who have mastered know-how of experimental tests is the coincidence of the results obtained in different trials of the same experiment. That is why in the case of any inconsistency of the results the experimenter usually does not satisfy himself with rejecting the result which is not confirmed by subsequent trials or rejecting the result which does not conform to the theory but tries to learn what has been responsible for arriving at the wrong, or apparently wrong result, in order to be able to avoid divergency of results in the future.

An experiment which in all trials brings the same result will be referred to as *dispersion free*, or else *dispersive*. In view of the remarks made above non-dispersiveness seems to be an attribute of fundamental significance. And in fact it is, but with a qualification.

When a postulated degree of accuracy ε in testing α is high, non-dispersiveness may become unavailable. The parameters which do not affect the result of experiment when ε has been taken sufficiently large, may become of considerable significance when ε is small. The slightest stirring of air may effect the results of measurements carried out with a balance weight. The slightest changes in humidity of air may change data concerning the characteristic of an electric system (say, capacities of condensers). Factors which are "normally" neutral become of crucial importance.

Often the influence of "undesirable" parameters on the result of an experiment may be at least to some extent minimized or removed. Very sensitive balance weight may be placed under glass casing with the air pumped out. The humidity of air may be under constant control. A schema of an experiment may be useless unless accompanied by a detailed instruction how to implement it in practice.

Even if in spite of all precaution measures the experimenter is not able to eliminate the sources of dispersiveness of the experiment, it does not mean that the data obtained in subsequent trials have no empirical value. Being inconsistent they still may form sequences in which relative frequency of some data (or some configurations of data) is in a sufficiently long series constant. The results of a sufficiently large amount of trials of the experiment may be then interpreted in terms of probability.

In the situation described one should redefine the objective of the experiment. Instead of treating the experiment as concerned with α we should start treating it as an experiment which is to decide what the value of $P(\alpha)$ is, where P stands for probability.

Furthermore, one should redefine the schema of dispersive experiment for α so as to arrive at a dispersion free experiment for sentences of the form $P(\alpha) = x$, or rather its approximations of the form $P(\alpha) = x \pm \varepsilon$.

Although the main idea of transforming a dispersive experiment concerning α into a dispersion free experiment concerning $P(\alpha) = x$ is clear, I believe, still the notation we have applied is highly misleading. The symbol P must be furnished with a subscript which indicates the way in which the probability P is measured. To see why it is so important consider the following.

Suppose that there are two different experimental schemas σ_1, σ_2 for α, neither of them being dispersion free, but both capable of being transformed into dispersion free schemas σ_1^P, σ_2^P for $P(\alpha) = x$. But the probability of α measured with the help of σ_1^P need not be the same as that measured with the help of σ_2^P. The reasons for that are obvious. Imagine any two measuring instruments, say two volt-meters. Let the accuracy of one of them be 10^{-3} V and that of the second 10^{-4} V. Applying the latter volt-meter the experimenter may measure the voltage in the dispersion free way in the case when the use of the former has already resulted in dispersive data. Clearly then the estimation of probability of α may depend on equipment involved in the experiment (say on selecting more or less accurate volt-meter). Note now that α_1 and α_2 may involve measurements of quite a different sort. The quantities which must be measured when an experiment falls under the schema α_1 need not be relevant for experiments based on α_2 and vice versa.

The fact that the value of $P(\alpha)$ may depend on the way in which P is measured indicates that we do not deal with one quantity "probability" but rather with a bunch of probability measures $P_1, P_2, \ldots, P_k$, all of them being "objective" in the sense that all of them can be measured by means of a dispersion free experimental procedure. One of the problems which we are going to discuss is how the empirical evidence presented by means of hypothesis of the form $P_i(\alpha) = x$ can be exploited both to examine the soundness of empirical theories and to contribute to defining empirical systems which are to represent the phenomena examined.

The notion of an experimental scheme allows us to be a bit more specific on what is to be meant by an experimental procedure. Whenever the notion of an experimental procedure is applied it will be understood that the experimental procedures referred to are defined by suitable experimental schemas and the instruments by means of which the experiments are carried out.

Given a fixed set of procedures τ and given a particular phenomenon p to which the procedures in τ can be applied, we shall denote by H_p the set of all hypotheses concerning p which are decidable by means of τ, and by τ_p the probability measure defined on H_p by the procedures in τ.

4.5. NOTES

A. *The role of concatenation operation in measurement.* Starting from N. R. Campbell, whose works on measurement (1920, 1928) influenced considerably many later philosophers (e.g. M. R. Cohen and E. Nagel 1934), it has often been taken for granted that one of the basic concepts indispensable in any analysis of measurement is that of concatenation of objects. In our exposition of some selected parts of measurement theory we mentioned explicitly the concatenation operation only in a few cases, notably when discussing the measurement of mass (Section 4.2, Example 2). The reader should not jump to the conclusion that the intention underlying the approach accepted was to diminish the role of the concatenation operation.

The motivation for the choice of the way of exposing the problem and for the selection of results of measurement theory presented in this book is two-fold. First, from the dydactic point of view (the conciseness and legibility of the exposition), selection of the interval ordering instead of the concatenation operation O seems to be a better idea. After all when considering the possibility of defining a numerical representation of a certain attribute (say brightness of the source of light, pitch of sound, level of intelligence, or strength of wind) we try to invent a scale which allows us to speak about difference among the objects on which the attribute is measured in a standarized and well defined manner.

The exposition of measurement theory in which the interval ordering is selected to play the key concept seems to reflect better the intuitive way of thinking about the problems of measurement.

Still there is another reason why the interval ordering seems to be a better choice than that of the concatenation relation. As it was convincingly argued by many authors (cf. e.g. T. W. Reese, 1943) looking for a suitable concatenation operation need not be the best approach, or even need not be the proper one, to the problem of defining a scale for measuring a magnitude. The relation structure involved in a measurement need not contain the concatenation operation. If it does, it defines a measurement structure of the sort we have discussed, i.e. comprising the interval relation. The converse need not be true. Thus the approach adopted here is more general as it covers, although implicitly, all measurements based on concatenation operation, and some others in addition.

I am far from claiming however that this approach is fully general. It may certainly happen that given a magnitude and a scale f in which it is measured, we are not able to find any empirical interpretation to the differences of the form $|f(a)-f(b)|$. Considerations of this sort reinforced by suitably selected examples provide a justification for a fully general and fully abstract approach to the problems of measurement theory (D. Scott and P. Suppes, 1958). The excerpt from D. H. Krantz, R. D. Luce, P. Suppes and A. Tversky, 1971, pp. 8,9 quoted below presents a very concise exposition of such an approach.

"Given an empirical relation R on a set A and a numerical relation S on Re, a function Φ from A into Re takes R into S provided that the elements $a, b, \ldots$ in A stand in relation R if and only if the corresponding numbers $\Phi(a), \Phi(b), \ldots$ stand in relation S. More generally, if $\langle A, R_1, \ldots, R_m \rangle$ is an empirical relational structure and $\langle Re, S_1, \ldots, S_m \rangle$ is a numerical relational structure, a real-valued function Φ on A is a homomorphism if it takes each R_i into S_i, $i = 1, \ldots, m$. Still more generally, we may have n sets $A_1, \ldots, A_n$, m relations $R_1, \ldots, R_m$ on $A_1 \times \ldots \times A_n$, and a vector-valued homomorphism Φ, whose components consist of n real-valued functions $\Phi_1, \ldots, \Phi_n$ with Φ_j defined on A_j, such that Φ takes each R_i into relation S_i on Re^n. A representation theorem asserts that if a given relational structure satisfies certain axioms, then a homomorphism into a certain numerical relational structure can be constructed. A homomorphism into the real numbers is often referred to as a scale in the psychological measurement literature.

From this standpoint, measurement may be regarded as the construction of homomorphism (scales) from empirical relational structures of interest into numerical relational structures that are useful. Foundational analysis consists, in part, of clarifying (in the sense of axiomatizing) assumptions of such constructions."

B. *Terminology.* The reader should be warned that the exposition of measurement theory provided in this book involves concepts which cannot be found elsewhere and, which follows, also names especially coined for them.

Thus for instance the concept of an interval structure is not applied in the way we have done. Still, it is a variant of similar concepts which are known. I mean here various sorts of the so-called *difference structures* (positive difference structures, algebraic-difference structures, absolute-difference structures, finite, equally spaced difference structures and others). For an exhaustive encyclopaedia of those structures cf. D. H. Krantz, R. D. Luce, P. Suppes and A. Tverski, 1971, Chapter 4.

In spite of all terminological nuances of the sort mentioned above, the exposition of measurement theory presented in this book does not deviate in any essential way from the main ideas on which this theory is based.

V. OPERATIONAL STRUCTURES

5.0. INTRODUCTORY ASSUMPTIONS

Considerations carried out in this chapter will refer to a fixed empirical theory $\Theta_{\mathscr{K}}$, Θ being its deductive part, and $\mathscr{K}$ its scope. We shall assume that $\Theta_{\mathscr{K}}$ satisfies the conventions listed below.

CONVENTION 5.0.1a. The language $\mathscr{L}_\Theta$ of $\Theta_{\mathscr{K}}$ is a standard semi--interpreted language. In particular then, given any standard interpretation I of $\mathscr{L}_\Theta$, the set of sentences valid under the interpretation I, $Val(I)$, is defined.

CONVENTION 5.0.1b. Call an extension $\mathscr{L}_\Theta^*$ of $\mathscr{L}_\Theta$ *inessential* if it results from $\mathscr{L}_\Theta$ by enlarging the vocabulary of the latter by constant symbols for real numbers or sets of real numbers. In what follows we shall not discern between $\mathscr{L}_\Theta$ and its inessential extensions.

* *
*

The last convention enables us to treat all sentences of the form $F_i(t) = x$ as well as all sentences of the form $F_i(t) \in \Delta$ as elements of $\mathscr{L}_\Theta$. Indeed it is enough to replace t, x and Δ by constant symbols and to fix interpretation for them in order to form an inessential extension of $\mathscr{L}_\Theta$ and thus to form a language which by Convention 5.0.1b is indiscernible from the original one.

* *
*

CONVENTION 5.0.2. $u, F_1, \ldots, F_n, G_1, \ldots, G_m$ are all specific constant symbols of $\mathscr{L}_\Theta$. The symbol u is assumed to denote an interval of real numbers (meant to represent an interval of time). Both F_i and G_j are functions from u into Re, i.e. under an empirical interpretation they are viewed to be quantities depending on time and taking real numbers as their values.

* *
*

Given an elementary system of the form

(1) $\mathfrak{a} = (u, F_1^a, \ldots, F_n^a, G_1^a, \ldots, G_m^a)$,

we shall call it a *partial model structure for* $\mathscr{L}_\Theta$ provided that there exists a standard interpretation I for $\mathscr{L}_\Theta$ such that $\mathfrak{a}$ is the restriction of I to the specific terms of $\mathscr{L}_\Theta$ which satisfies the conditions: $I(u) = T(\mathfrak{a})$, $I(F_i) = F_i^a$, $i = 1, \ldots, n$, and $I(G_j) = G_j^a$, $j = 1, \ldots, m$. Note that if $\mathfrak{a}$ is a partial moder structure for $\mathscr{L}_\Theta$, then by Convention 5.0.1 and by the definition of a standard semi-interpreted language (cf. 1.8) there exists exactly one standard interpretation I such that $\mathfrak{a}$ is the restriction of I to the specific terms of $\mathscr{L}_\Theta$.

The reduct of $\mathfrak{a}$ to $F_1^a, \ldots, F_n^a$ will be referred to as ***F-reduct*** *of* $\mathfrak{a}$ and denoted as $\mathfrak{a}^F$.

* *
*

CONVENTION 5.0.3. There exists a fixed set of empirical procedures τ by means of which empirical hypotheses relevant for $\Theta_{\mathscr{K}}$ are decided.

* *
*

In order to call attention to the fact that τ are the procedures applied in $\Theta_{\mathscr{K}}$ we shall occasionally denote this theory as $\Theta_{\mathscr{K}}^\tau$. Note that $\Theta_{\mathscr{K}}^\tau$ can be viewed as the triple $(\Theta, \tau, \mathscr{K})$ consisting of a deductive theory Θ (the content of $\Theta_{\mathscr{K}}^\tau$), a range of applicability $\mathscr{K}$ of Θ (the scope of $\Theta_{\mathscr{K}}^\tau$) and a set τ of testing procedures. We shall occasionally refer to $\Theta_{\mathscr{K}}^\tau$ or equivalently to $(\Theta, \tau, \mathscr{K})$ as to a *theory in methodological sense.*

Let us denote by $\boldsymbol{B}(Re)$ the family of all Borel sets on the real line. As known, $\boldsymbol{B}(Re)$ is the least σ-algebra of sets generated by closed intervals of Re, i.e. is the least family of subsets of Re which satisfies the following conditions:

(B1) If Δ is a closed interval of real numbers, then Δ is in $\boldsymbol{B}(Re)$

(B2) If Δ is in $\boldsymbol{B}(Re)$, then $-\Delta = Re - \Delta$ is in $\boldsymbol{B}(Re)$

(B3) If Δ_i, $i = 1, 2, \ldots$, are in $\boldsymbol{B}(Re)$, then $\bigcup \{\Delta_i : i \in \mathcal{N}\}$ is in $\boldsymbol{B}(Re)$.

With the help of (B1)-(B3) one may easily prove that the following conditions are satisfied:

(B4) $Re \in \boldsymbol{B}(Re)$.

(B5) $\varnothing \in \boldsymbol{B}(Re)$.

(B6) For each $x \in Re$, $[x, x] \in \boldsymbol{B}(Re)$.

(B7) For each $x, y \in Re$, if $x < y$, then $(x, y]$, $[x, y)$, (x, y) are in $\boldsymbol{B}(Re)$.

(B8) If Δ_1, Δ_2 are in $\boldsymbol{B}(Re)$, so is $\Delta_1 - \Delta_2$.

(B9) If Δ_i, $i = 1, 2, \ldots$ are in $\boldsymbol{B}(Re)$, then also $\bigcap \{\Delta_i : i \in \mathcal{N}\}$ is in $\boldsymbol{B}(Re)$.

In what follows the symbol Δ, provided with subscripts or superscripts if necessary, will always refer to a *Borel set.*

* *
*

CONVENTION 5.0.4. $F_1, \ldots, F_n$ are all specific terms of $\mathcal{L}_\Theta$ which appear in τ-decidable hypotheses, i.e. decidable by means of the procedures τ. On the other hand, for each p in $\mathcal{K}$ and for each F_i, there exist t and Δ such that the sentence $F_i(t) \in \Delta$ is τ-decidable.

* *
*

Note that under this convention $F_1, \ldots, F_n$ are *empirical* terms of $\Theta_{\mathcal{K}}$. All the remaining terms (cf. the terminology adopted in Section 3.4) will be referred to as *theoretical.*

* *
*

CONVENTION 5.0.5. In what follows the symbol F will represent any of the symbols $F_1, \ldots, F_n$. Thus, for instance, by a sentence of the form $F(t) = x$ we shall mean any sentence of the form $F_i(t) = x$, $i = 1, \ldots, n$ (needless to say, in order for $F_i(t) = x$ to be a sentence, t and x should be constant symbols).

CONVENTION 5.0.6. We shall often write (F, t, Δ) instead of $F(t) \in \Delta$, and (F, t, x) instead of $F(t) \in [x, x]$ (or equivalently $F(t) = x$).

* *
*

Let us conclude this section by the following, as a matter of fact quite obvious, remark. In order for a sentence of the form (F, t, Δ) to have a definite empirical meaning, a scale for measuring F and similarly for measuring time should be selected. Unless it has been done, t does not refer to any specific instant and does not indicate any specific interval of values of F. For an example consider the formula $F_1(0) = 5.2 \pm 0.1$ (or equivalently $(F_1, 0, [5.2-0.1, 5.2+0.1])$). Although from the mathematical point of view it is perfectly well formed, from the empirical point of view it is ambiguous unless we know how the scale of F_1 has been defined (i.e. what has been selected to be the unit in which the values of F_1 are measured and perhaps also what serves as the zero point of the scale for F_1 or even what forms the frame of reference with respect to which F_1 is measured).

We shall assume that given any p in $\mathscr{K}$, the scales of time they both will be applied only in informal comments.

5.1. VERIFICATION PROCEDURES

From the semantic standpoint, a sentence of the form (F, t, Δ) says that "F measured at time t has the value within Δ". The locution in the quotation marks will be occasionally referred to as the *objective meaning of* (F, t, Δ). In contradistinction to the objective meaning of (F, t, Δ), by the *operational meaning* of that sentence we shall understand the set of procedures in τ by means of which (F, t, Δ) can be decided, Neither the objective meaning nor the operational one should be treated as technical notions, they both will be applied only in informal comments.

Note that under the objective interpretation it makes sense to think of (F, t, Δ) as of a symbol composed of the elements F, t, Δ, each of them being meaningful all by itself, i.e. F standing for a particular quantity, t for a particular instant, and Δ for a particular Borel set. Under the operational interpretation the symbols F, t, Δ considered outside the context of the sentence (F, t, Δ) have in fact no meaning at all. At least they have no meaning unless the value $F(t)$ of F at t can be measured in a precise way. If an approximate measurement of F is possible, only the procedures in τ define an approximate measure for F, not F itself. Furthermore, the very existence of F must be treated as an (operationally) meaningless hypothesis.

Similar remarks apply to the symbol t. There are no empirical procedures by means of which time can be measured in a precise way. Thus, only an approximate measure of time is operationally defined, not time itself, although it is both customary and convenient to speak about measuring F at time t. In order to perform the measurement operation and to decide whether (F, t, Δ) or not, it is necessary to assume neither the existence of a "sharp-cut" quantity F nor the existence of a "sharp-cut" quantity t.

The following obvious remark should complement our comments on the meaning of (F, t, Δ). In the case of both objective and operational interpretations of that hypothesis, before one starts considering what (F, t, Δ) means, one must point out a particular domain p in $\mathscr{K}$ (or at least a particular possible domain of Θ) to which (F, t, Δ) is referred. The results of measuring F in an application p need not coincide with those referring to q. We should write then (F^p, t, Δ) and (F^q, t, Δ) respectively, rather than (F, t, Δ). Yet, the latter notation is not only simpler but also proper if (F, t, Δ) is to be considered as a formula of $\mathscr{L}_\Theta$. Symbols of the form F^p will be occasionally used in metalinguistic comments, but they do not belong to the language $\mathscr{L}_\Theta$, and thus to inessential extensions of it either.

Our nearest task will be to define the notion of a verification function and impose certain conditions on testing procedures τ. However, before we do that let us define a certain class H of sentences of $\mathscr{L}_\Theta$; we shall call them experimental-like.

DEFINITION 5.1.1. *The set H of all* experimental-like hypotheses *of $\mathscr{L}_\Theta$ is the least set of sentences of $\mathscr{L}_\Theta$ that satisfies the following two conditions*:

(i) *Each sentence of the form (F, t, Δ) is in H.*

(ii) *If α, β are in H, then $\neg\alpha, \alpha \vee \beta$, and $\alpha \wedge \beta$ are in H.*

The definition requires a few words of justification. As one may easily guess, by an experimental-like sentence we want to understand such a sentence which is a candidate for being experimentally decidable. This motivates including into H the sentences of the form (F, t, Δ), with Δ being a proper interval of real numbers. Still some formal considerations suggest it desirable to have all elementary sentences in H, i.e. to admit all Borel sets as Δ, although some of sentences of the form (F, t, Δ) might be by their very nature completely inaccessible to any empirical verifications.

Sentences of the form $(F, t, \varnothing)$ and (F, t, Re) are of special sort among all experimental-like sentences. $(F, t, \varnothing)$ is mathematically inconsistent, i.e. its negation is mathematically provable. On the other hand, all sentences of the form (F, t, Re) are mathematical identities.

The decision to include into H "molecular" sentences, i.e. sentences of the form $\neg\alpha, \alpha \vee \beta, \alpha \wedge \beta$ needs some comment. Clearly one may easily figure out tests which are applicable to compound sentences (say of the form $\alpha \vee \beta$) though they are not applicable to their components. This observation provides a partial justification for Definition 5.1.1. Much more important reason for considering molecular sentences as experimental-like is that in the case of dispersive procedures it is not enough to evaluate probabilities of elementary sentences in order to calculate the probability of sentences composed of them, say their conjunctions. If α and β are deductively independent, the fact that the values of $P(\alpha)$ and $P(\beta)$ have been established may not suffice to compute $P(\alpha \wedge \beta)$. Thus to establish the value of $P(\alpha \wedge \beta)$ (which incidentally allows to compute the conditional probability $P(\alpha/\beta)$) some additional experiments concerning just the conjunction $\alpha \wedge \beta$ may happen to be necessary.

The sentences in H are not independent. Note however that it is not due to the fact that H involves compound hypotheses. The set-theoretical relations which Borels sets bear among themselves entail both semantic and deductive dependences among experimental-like sentences. For instance (F, t, Δ) implies $(F, t, \Delta \cup \Delta')$. It also implies not $(F, t, -\Delta)$. More subtle interdependences among experimental-like sentences can be due to some special conditions imposed on standart interpretations for F. For instance, if F is required to be continuous under any standard interpretation, some infinite sets of experimental-like sentences should be semantically inconsistent (have no standard interpretation) though their finite subsets are not.

Write (cf. Section 1.11) $\alpha \sim \beta$ if both $\alpha \vdash \beta$ and $\beta \vdash \alpha$.

ASSERTION 5.1.2. *The structure* $(H/\sim, \cup, \cap, -)$, *with* $\cup, \cap$ *being binary operations and* $-$ *a unary operation on* $H/\sim$ *defined by*

(i) $|\alpha| \cap |\beta| = |\alpha \wedge \beta|$,

(ii) $|\alpha| \cup |\beta| = |\alpha \vee \beta|$,

(iii) $\neg |\alpha| = |\neg \alpha|$,

is a Boolean algebra, $|\alpha \wedge \neg \alpha|$ *being the least and* $|\alpha \vee \neg \alpha|$ *the greatest element of it.*

PROOF. $(H/\sim, \cup, \cap, -)$ is a particular example of the Lindenbaum-Tarski algebra, and thus 5.1.2 is an immediate corollary from a more general result (cf. Theorem 1.11.5).

DEFINITION 5.1.3.

A. *A function* $v: H_v \to \{0, 1\}$, *where* $H_v \subseteq H$, *will be said to be a* verification function *if and only if the following two conditions are satisfied*:

(v1) *There exists a Boolean homomorphism* V *from the Lindenbaum-Tarski algebra* $(H/\sim, \cup, \cap, -)$ *onto the two element Boolean-algebra* $(\{0,1\}, \cup, \cap, -)$ *such that* $V(|\alpha|) = v(\alpha)$ *for all* α *in* H_v.

(v2) *If* $\alpha_1, \ldots, \alpha_n$ *are in* H_v, β *is in* H *and there exists* $v \in \{0, 1\}$ *such that for each Boolean homomorphism* V *from* $(H/\sim, \cup, \cap, -)$ *onto* $(\{0, 1\}, \cup, \cap, -)$, $V(|\beta|) = v$ *whenever*

$$V(|\alpha_1|) = \mathrm{v}(\alpha_1), \ldots, V(|\alpha_n|) = \mathrm{v}(\alpha_n),$$

then β *is in* H_v *and* $\mathrm{v}(\beta) = v$.

B. *If moreover*

(v3) *If* $X \models \beta$, $X \subseteq H_v$, $\beta \in H$ *and* $\mathrm{v}(\alpha) = 1$ *for all* α *in* X, *then* β *is in* H_v *and* $\mathrm{v}(\beta) = 1$,

then v *is said to be* semantically admissible.

As we are going to prove, Definition 5.1.3 can be easily couched in a simpler language, nevertheless we intentionally decided to utter it in its quite sophisticated form. Worded as it has been done, Definition 5.1.3 is a ready analogue of Definition 5.6.2, the latter referring to testing functions of a more complicated sort than verifications.

The reader may easily verify that

Assertion 5.1.4. *If* v *is a verification function, the following conditions are satisfied*:

(v4) $\mathrm{v}(\alpha) = 1$, *for all* α *in* H_v *such that* $\vdash \alpha$.

(v5) $\mathrm{v}(\alpha) = 0$, *for all* α *in* H *such that* $\vdash \neg\alpha$.

(v6) *If* α, β *are in* H_v *and* $\alpha \sim \beta$ *then* $\mathrm{v}(\alpha) = \mathrm{v}(\beta)$.

(v7) α *is in* H_v *if and only if* $\neg\alpha$ *is in* H_v.

(v8) *If* $\alpha_1, \ldots, \alpha_k \vdash \beta$, $\alpha_1, \ldots, \alpha_k$ *are in* H_v, β *is in* H *and* $\mathrm{v}(\alpha_1) = \ldots = \mathrm{v}(\alpha_k) = 1$, *then* β *is in* H_v, *and* $\mathrm{v}(\beta) = 1$.

Let us put

$$E_v = \{\alpha \in H_v : \mathrm{v}(\alpha) = 1\}.$$

Assertion 5.1.5. *Given any sentence* α *in* H_v (H_v *being the domain of the verification function* v) *exactly one of the sentences* α *or* $\neg\alpha$ *is in* E_v.

Proof. If $\alpha \in H_v$ and $\mathrm{v}(\alpha) = v$ then, by (v1), $V(|\alpha|) = v$ and thus $V(|\neg\alpha|) = -v$. Since there is no choice in selecting the value of $V(|\neg\alpha|)$, therefore, by (v2), α must be in H_v and moreover we must have $\mathrm{v}(\neg\alpha) = -v$. Thus, depending on whether $v = 1$

or $v = 0$, $\alpha \in E_v$ or $\neg\alpha \in E_v$ but not both. If $\neg\alpha \in H_v$, the argument runs in analogous way.

COROLLARY 5.1.6. $E_v = \{a \in H_v\colon v(\alpha) = 1 \text{ or } v(\neg\alpha) = 0\}$.

The simple facts we have established thus far allow us to prove:

ASSERTION 5.1.7. *If v is a verification function, the set E_v is deductively closed in H, i.e. for each α in H, $\alpha \in E_v$ whenever $E_v \vdash \alpha$.*

PROOF. Suppose that $\alpha \in H$ and $E_v \vdash \alpha$. The derivability relation is finitary and hence for some $\alpha_1, \ldots, \alpha_n$ in E_v, $\alpha_1, \ldots, \alpha_n \vdash \alpha$. We have $v(\alpha_1) = \ldots = v(\alpha_n) = 1$ and thus by ($v8$), α is in H_v and $v(\alpha) = 1$. Thus $\alpha \in E_v$ which concludes the proof.

Let us comment briefly on this part of the proof in which we make use of condition ($v8$). Since the proof of that condition has not been given it is perhaps worth-while mentioning that ($v8$) is an easy corollary from ($v2$). Indeed if $v(\alpha_1) = \ldots = v(\alpha_n) = 1$ and $\alpha_1, \ldots, \alpha_n \vdash \beta$, then for each Boolean homomorphism V such that $V(|\alpha_i|) = v(\alpha_i)$, $i = 1, \ldots, n$, $V(|\beta|)$ must be equal 1. Hence $v(\beta) = 1$.

Assertions 5.1.5 and 5.1.7 entail:

COROLLARY 5.1.8. *If v is a verification function, E_v is deductively consistent.*

The following holds true:

ASSERTION 5.1.9. *A function $v\colon H_v \to \{0, 1\}$, where $H_v \subseteq H$, is a verification function if and only if the following two conditions are satisfied:*

(i) *the set*

$$E_v = \{\alpha \in H_v : v(\alpha) = 1\}$$

is

a. *deductively consistent, and*

b. *deductively closed in H.*

(ii) $H_v = E_v \cup \{\alpha\colon \neg\alpha \in E_v\}$.

PROOF. We have already proved the "only if" part of the assertion. Suppose now that E_v is both deductively consistent and deductively closed in H_v. By Theorem 1.11.6, the deductive consistency of E_v guarantees that there exists a homomorphism V from $H/\sim$ onto the two element Boolean algebra such that $V(|\alpha|) = 1$ for all $\alpha \in E_v$.

Suppose that $\alpha \in H_v - E_v$, then $v(\alpha) = 0$. But at the same time by (ii) $\neg\alpha \in E_v$ and thus $V(|\alpha|) = 0$, which shows that condition ($v1$) is satisfied.

In order to prove ($v2$) assume that $\alpha_1, \ldots, \alpha_n$ are in H_v, and consider sentences $\alpha'_1, \ldots, \alpha'_n$, where $\alpha'_i = \alpha_i$, provided that $v(\alpha_i) = 1$ or else $\alpha'_i = \neg\alpha_i$. Clearly $\alpha'_1, \ldots, \alpha'_n$ are again in H_v, and $v(\alpha'_1) = 1$ for all $i = 1, \ldots, n$. Suppose that for each Boolean homomorphism V such that

(1) $$V(|\alpha_i|) = v(\alpha_i), \quad i = 1, \ldots, n$$

$V(|\beta|)$ has a fixed value v. Put $\beta' = \beta$ when $v = 1$ or else $\beta' = \neg\beta$. Observe that for each V, $V(|\beta'|) = 1$ whenever

(2) $$V(|\alpha'_1|) = \ldots = V(|\alpha'_n|) = 1.$$

This proves that for all V

$$V(|\alpha'_1 \wedge \ldots \wedge \alpha'_n \rightarrow \beta'|) = 1,$$

which, by Theorem 1.11.6 and by Deduction Theorem (cf. Assertion 1.10.3), proves that

$$\alpha'_1 \wedge \ldots \wedge \alpha'_n \vdash \beta'.$$

Since $\alpha'_1, \ldots, \alpha'_n$ are in E_v and E_v is deductively closed, then β' is in E_v. This yields $v(\beta') = V(|\beta'|) = 1$, for each V such that (2) is satisfied. Consequently $v(\beta) = V(|\beta|)$ for each V such that (1), which concludes the proof.

As the reader has already noticed, Assertion 5.1.9 is an alternative (and simpler) definition of a verification function. Let us now briefly comment on the notion of admissibility of a verification function (cf. part B of Definition 5.1.3).

In general E_v need not be semantically consistent. To see that assume that H_v is the least set which comprises all sentences of the form (F_1, t_0, Δ), (t_0 being a fixed instant) and is closed under forming negations, disjunctions and conjunctions. In turn define v as a verification function which Desides Conditions (v1)–(v2) of Definition 5.1.3 satisfies the following one

(*) $v(F_1, t_0, \Delta) = 1$ and only if for some $n > 0, (0, 1/n] \subseteq \Delta$.

Denote by E^0 the set of all sentences of the form (F_1, t_0, Δ) such that $v(F_1, t, \Delta) = 1$. One may easily verify that, if no special mathematical postulates are imposed on F_1, E_v^0 is consistent. Indeed, select any finite subset $(F_1, t_0, \Delta_1), \ldots, (F_1, t_0, \Delta_n)$ of E_v^0. Then such a subset is deductively equivalent to

(3) $(F_1, t_0, \Delta_1 \cap \Delta_2 \cap \ldots \cap \Delta_n)$

and since $\Delta_1 \cap \Delta_2 \cap \ldots \cap \Delta_n$ is easily seen to be not empty, (3) has a semantic model, and it must be consistent, unless certain special assumptions concerning F_1 rule out this model from the class of standard models for F_1. In that way we have proved that all finite subsets of E_v^0 may be treated as consistent which, in view of the finitary character of $\vdash$, yields consistency of E_v^0.

On the other hand, one may prove that E_v must coincide with $\{\alpha \in H_v : E_v^0 \vdash \alpha\}$. We shall omit this part of argument because the technicalities in involves are of minor significance for further considerations. The consistency of E_v^0 implies then the consistency of E_v.

Being deductively consistent the set E_v^0 is semantically inconsistent for an obvious reason. There does not exist any standard interpretation in which all sentences of the form $(F_1, t_0, (0, 1/n])$ are simultaneously satisfied since the intersection of all intervals of the form $(0, 1/n]$ is empty!

The semantic consistency of E_v is guaranteed only by semantic admissibility of the verification function v.

ASSERTION 5.1.10. *E_v is semantically consistent if it is semantically admissible.*

The proof of this assertion is straightforward.

5.2. DISPERSION FREE OPERATIONAL STRUCTURES

Let us assume that, for each p in $\mathscr{K}$, the function τ_p (i.e. the function defined by the testing procedures τ of $\Theta^{\tau}_{\mathscr{K}}$) is a verification function. Abbreviate H_{τ_p} as H_p and E_{τ_p} as E_p. Each sentence α in H_p will be referred to as *τ-decidable*; if moreover it is in E_p, it will be referred as *τ-valid*.

Whatever p may be under the assumptions we made, the union $E_p \cup \Theta$ represents all we may learn about p with the help of both the laws and empirical procedures of the theory Θ. The following two definitions have then an obvious motivation.

DEFINITION 5.2.1. *An elementary system* $\mathfrak{a}_p = (T(p), F_1^p, \ldots, F_n^p)$ *is an* ideal model of p permitted by τ *if and only if all sentences in* E_p *are valid in* $\mathfrak{a}_p$.

DEFINITION 5.2.2. *The theory* $(\Theta, \tau, \mathscr{K})$ *is* deductively (semantically) sound *in* $p \in \mathscr{K}$ *if and only if* $\Theta \cup E_p$ *is deductively (semantically) consistent.*

The concept of soundness as it is defined above depends on testing procedures and in that sense is (partially at least) operational. In order to have a purely semantic definition of soundness of $\Theta_{\mathscr{K}}$ in p, we must imitate in a general manner the procedure exemplified by considerations of Chapter III, Section 3.5. Namely, we must assume that p is represented by a system $\mathfrak{a}_p$ and that there is a class $c(\mathfrak{a}_p)$ of standard interpretations for $\mathscr{L}_\Theta$ corresponding to $\mathfrak{a}_p$. The next step would consist in defining $\Theta_{\mathscr{K}}$ to be sound in p if and only if for each $\mathfrak{a} \in c(\mathfrak{a}_p)$, $\Theta \subseteq Val(\mathfrak{a})$. This is exactly the way in which Definition 3.5.1 was formed.

As long as we do not worry how exactly $\mathfrak{a}_p$ is to be defined there is no trouble with implementing the program presented above. Since phenomena p in $\mathscr{K}$ are examined in terms of variables $F_1, \ldots, F_n$ it is natural to assume that $\mathfrak{a}_p$ is an elementary system of the form $(T(p), F_1^p, \ldots, F_n^p)$. $c(\mathfrak{a}_p)$ should be then defined as the set of all partial model structures $\mathfrak{a}$ for $\mathscr{L}_\Theta$ such that $\mathfrak{a}^F = \mathfrak{a}_p$.

The belief that each singular phenomena in the scope of an empirical theory can be viewed as an elementary structure can be a perfectly workable assumption which does not lead to any inconsistency and provides a simple model of relations the theory bears to its domains. The assumption that p has an ideal model need not contradict either our theoretical knowledge or empirical evidence available. Furthermore, one can speculate about "potential" definability of $\mathfrak{a}_p$ by successive improvements of testing procedures.

Given two verification functions v, v', both being applicable to a domain p of $\Theta_{\mathscr{K}}$, define v' to be *more accurate* than v, whenever $E_p^v \subset E_p^{v'}$, i.e. each sentence v-decidable is v'-decidable and moreover all sentences v-valid are v'-valid.

The theory $(\Theta, \tau, \mathscr{K})$ may be treated as an initial one in the sequence of theories

$$(\Theta, \tau, \mathscr{K}), (\Theta', \tau', \mathscr{K}'), \ldots,$$

each theory in the sequence being an improvement of the one preceding it. The notion of an improvement involved in this remark need not be clear, but what concerns us for the time being is that the accuracy of verification functions

(1) $\tau_p, \tau'_p, \tau''_p, \ldots$

defined by the procedures $\tau, \tau', \tau'', \ldots$ respectively may increase, and we may speculate that there is exactly one system $\mathfrak{a}_p$ permitted by all verification functions forming the sequence (1). Denote by $Id^i(p)$ the set of ideal models of p permitted by τ^i. If τ'_p is more accurate than τ_p, τ''_p is more accurate than τ'_p, and so forth, then we clearly have

(2) $\ldots Id''(p) \subset Id'(p) \subset Id(p).$

The assumption that there is exactly one model $\mathfrak{a}_p$ permitted by all elements of $\mathfrak{a}_p$ amounts to the assumption that the intersection

(3) $Id(p) \cap Id'(p) \cap Id''(p) \cap \ldots$

coincides with $\{\mathfrak{a}_p\}$.

Regardless of how appealing the idea of an ideal model $\mathfrak{a}_p$ of p is, there are good reasons to doubt whether unlimited accuracy of measurement is available and thus whether it makes sense to expect that intersection (3) may turn out to be a unit set. In connection with this remark let me quote the following excerpt from M. I. Dalla Chiara Scabia and F. Toraldo di Francia's discussion on the notion a physical quantity (cf. 1973, p. 8):

"The purpose of a measurement allegedly is to attach to a physical quantity Q in a case s under examination a number q, called the value. Is this a real number, say with an infinity of decimal places?

It is everyday experience of the scientist working in a laboratory that the result of a measurement can never be a real number. It is instead an interval of real numbers. If the length of the interval is ε and q its midpoint, it is customary to say that the result of the measurement is $q \pm \varepsilon/2$. The quantity ε multiplied by the unit of measurement represents the accuracy or precision of the procedure used. The smaller ε is, the more accurate measurement can be performed.

Sometimes we say conventionally that there is an error in the measurement and that the true value may be anywhere in the interval $q-\varepsilon/2$, $q+\varepsilon/2$. If taken literally, this locution is somewhat misleading, and can give rise to contradictions. It only means that if we now carry out the measurement with an instrument of any better precision $\varepsilon' < \varepsilon$, the new interval ε' can fall anywhere within the interval ε".

If the real valued quantities are to be treated as merely conventional invents which do not refer to anything existing in the real life, and still worse "can give rise to contradictions", we certainly have to reexamine the idea of representing empirical phenomena by abstract systems and either give it up completely or at least revise the approach we considered to be sound thus far.

If the latter option is to be chosen, what might be the alternative approach? The one we are going to present and discuss in a concise manner was proposed by M. I. Dalla Chiara Scabia and Toraldo di Francia (1973). (cf. also R. Wójcicki, 1974).

Call a system of the form

$$\mathbf{a} = (u, \Phi_1, \ldots, \Phi_n),$$

where u is a non-empty interval of real numbers and Φ_i are functions of the form

$$\Phi_i\colon u \to \boldsymbol{B}(Re),$$

an *operational dispersion free elementary system, df-system* for short. In what follows the interval u, which clearly plays the role of the duration of $\mathfrak{a}$, will be denoted as $T(\mathfrak{a})$, and as a rule will not be listed among the constituents of the system. In general by a *dispersion free operational system* we shall understand any system of the form

$$\mathbf{A} = (A, u, \Phi_1, \ldots, \Phi_n),$$

where A is a non-empty set, u (to be denoted also as $T(\mathbf{A})$) is an interval of real numbers and each system parameter Φ_i is a function of the form

$$\Phi_i : u \times A^{k(i)} \to B(Re).$$

Clearly there is a far going analogy between operational and numerical systems, and in particular between elementary operational and elementary numerical systems. Observe, in particular, that the concepts defined in order to examine numerical systems have their obvious operational counterparts.

For instance, given an elementary operational system $\mathbf{a} = (\Phi_1, \ldots, \Phi_n)$, the state $\mathbf{a}(t)$ of $\mathbf{a}$ at time $t \in T(\mathbf{a})$ should be defined to be the n-tuple

$$(\Phi_1(t), \ldots, \Phi_n(t)).$$

Tde function $\{\mathbf{a}(t)\}_{t \in T(\mathbf{a})}$ should be referred to as the *history of* $\mathbf{a}$, and so forth.

Note that elementary systems can be viewed as operational systems of special kind. If $\mathfrak{a} = (F_1, \ldots, F_n)$, $\mathbf{a} = (\Phi_1, \ldots, \Phi_n)$, $T(\mathfrak{a}) = T(\mathbf{a})$ and moreover for each $t \in T(\mathfrak{a})$, and for each Φ_i,

$$\Phi_i(t) = [F_i(t), F_i(t)],$$

$\mathfrak{a}$ is uniquely defined by $\mathbf{a}$ and vice versa. In what follows we shall treat such two systems as identical.

DEFINITION 5.2.3.

A. *A df-system* $\mathbf{a}_p^\tau = (\Phi_1^\tau, \ldots, \Phi_n^\tau)$ *will be said to be the* τ-model of the domain p of $\Theta^\tau_{\mathcal{X}}$ *if and only if the following conditions are satisfied*:

(01) $T(\mathbf{a}_p^\tau) = T(p)$.

(02) *For each* t *in* $T(p)$ *and for each* Φ_i,

$$\Phi_i^\tau(t) = \bigcap \{\Delta : \tau_p(F_i, t, \Delta) = 1,\ (F_i, t, \Delta) \in H_p\}.$$

B. *If moreover* $\Phi_i^\tau(t) \neq \emptyset$ *for each* t *in* $T(p)$ *and for each* Φ_i^τ, *the model* $\mathbf{a}_p^\tau$ *will be said to be* proper.

It is immediately seen that for each $p \in \mathscr{K}$ there exists exactly one τ-model $\mathbf{a}_p^\tau$ of p. However, it need not be proper. Still the following holds true.

ASSERTION 5.2.4. *If the verification function* τ_p *is semantically admissible, the* τ*-model* $\mathbf{a}_p^\tau$ *of* p *is proper.*

PROOF. Denote as Δ_0 the intersection which appears in Condition (03) of Definition 5.2.3., and denote as X_0 the set of all hypotheses (F_i, t, Δ) in H_p such that $\tau_p(F_i, t, \Delta) = 1$. We have $X_0 \models (F_i, t, \Delta_0)$ and hence, by (v3) of 5.1.3, $\tau_p(F_i, t, \Delta_0) = 1$. Since E_p is deductively consistent (cf. Corollary 5.1.8), $\Delta_0 \neq \emptyset$, which concludes the proof.

Note that the operational model $\mathbf{a}_p^\tau$ may be treated as if it were an "objective" model for p, i.e. one whose properties depend neither on theorems concerning p nor on testing procedures. One may pretend that the procedures in τ serve to discover properties of $\mathbf{a}_p^\tau$ and do not play definitional role. Although the decision to consider structures of the form $\mathbf{a}_p^\tau$ as objective (in the sense we have tried to characterize in a loose manner) models of phenomena in $\mathscr{K}$ would be somewhat artificial, still the idea to treat elements of $\mathscr{K}$ as operational systems deserves to be examined in more detail.

Let us turn back to our remarks concerning successive improvements $\tau, \tau', \tau'', \ldots$ of testing procedures being the result of constant progress in science. If the elements of the sequence $\tau, \tau', \tau'', \ldots$ are of increasing accuracy, we clearly have

$$\Phi_i^{\tau''}(t) \subseteq \Phi_i^{\tau'}(t) \subseteq \Phi_i^\tau(t).$$

Put

$$\Phi_i(t) = \Phi_i^\tau(t) \cap \Phi_i^{\tau'}(t) \cap \ldots$$

and define

$$\mathbf{a}_p = (T(p), \Phi_1, \ldots, \Phi_n).$$

As we have already mentioned, one should not expect that $\mathbf{a}_p$ would turn out to be an ideal model for p. What makes, however, the status of $\mathbf{a}_p$ very similar to that of the system $\mathfrak{a}_p$ discussed earlier is the fact that the two systems are viewed as limit systems whose successive, more and more accurate approximations are provided by the systems $\mathbf{a}_p^{\tau}, \mathbf{a}_p^{\tau'}, \mathbf{a}_p^{\tau''}, \ldots$. Similarly then to the case of $\mathfrak{a}_p$ one may claim $\mathbf{a}_p$ to be an "objective" operational dispersive (df-) model for p, i.e. the one which is indiscernible from p itself and which can be exhaustively examined only with the help of constant improvements of theoretical knowledge and experimental techniques. Since "objective operational model" is a sort of self-contradictory locution, in what follows we shall prefer to speak about *limit operational df-models.*

5.3. A REVISED NOTION OF REGULARITY

Let us start our considerations concerning the possibility to employ operational systems to model empirical phenomena with examining how this approach would affect the results of the discussion on regularities carried out in Chapter II.

One may easily see that all definitions stated in Chapter II may in an obvious way be rephrased for df-systems. For instance, given a set Z of similar operational systems (the notion of similarity is to be understood in a usual way), we may define Z to be *operationally state-determined* if and only if given any to systems a, b in Z, the identity

$$\mathbf{a}(t) = \mathbf{b}(s),$$

$t \in T(\mathbf{a}), s \in T(\mathbf{b})$, implies that

$$\mathbf{a}(t+r) = \mathbf{b}(s+r)$$

for all r such that $t+r \in T(\mathbf{a})$ and $s+r \in T(\mathbf{b})$.

The result of our short discussion is surprisingly simple but at the same time it need not be convincing. One may find the conclusion that the shift from numerical to operational structures requires merely suitable reinterpretation of the earlier definitions difficult to accept. He may argue as follows.

Although formally sound, the definition of a state-determined class of operational systems has no good factual interpretation.

Assume for instance that all systems in Z are ν-models of phenomena in $\mathscr{K}$, where ν is an arbitrary set of verification procedures (i.e. each test-function ν_p is a verification function). Put $Z = \{\mathbf{a}_p^\nu : p \in \mathscr{K}\}$ and assume that Z is state-determined. The system of the form $\mathbf{a}_p^\nu$ does not provide an objective representation of the phenomenon p but rather it represents p "as it looks like" when examined with the help of procedures in ν. If then

$$(1)^{\bullet} \qquad \mathbf{a}_p^\nu(t) = \mathbf{a}_q^\nu(t),$$

it does not suffice to draw the conclusion that in the "objective" sense p and q are at t at the same state. The fact that (1) holds true does not guarantee that

$$(2) \qquad \mathbf{a}_p^{\nu'}(t) = \mathbf{a}_q^{\nu'}(t),$$

for all procedures ν' we shall be ready to accept in the future.

Note however that the argument presented presupposes existence of procedures which define the systems $\mathbf{a}$ and $\mathbf{b}$, and furthermore it presupposes the possibility of further improvement of those procedures. But if someone is ready to take seriously the idea of representing phenomena by limit structures defined only by infinite sequences of subsequent approximations the argument fails.

When applied to limit operational models the definition of an operationally state-determined phenomenon Z is only apparently counterintuitive. It may be interpreted as saying that Z is operationally state-determined provided that whenever by no procedures whatsoever, i.e. neither by those applied nowadays nor by those we shall invent and accept in the future, it is possible to discern between $\mathbf{a}(t)$ and $\mathbf{b}(t)$, one is also unable to discern between $\mathbf{a}(t+r)$ and $\mathbf{b}(t+r)$ for any r. As a matter of fact this is exactly what the definition of state-determinism (of phenomena represented by elementary systems) amounts to when translated into the experimenter's language.

Though, perhaps, it is not immediately seen, the "ordinary" and the operational accounts of regularities are in a sense equivalent. To state this loose remark in a precise language let us introduce the notion of an idealization of an operational elementary system.

DEFINITION 5.3.1. *Let* $\mathbf{a} = (\Phi_1, \ldots, \Phi_n)$ *be a df-system. We shall say that an elementary system* $\mathfrak{a}$ *is an* idealization of $\mathbf{a}$, $\mathfrak{a} \in Id(\mathbf{a})$, *if and only if* $\mathfrak{a}$ *is a system of the form* $(F_1^{\mathfrak{a}}, \ldots, F_n^{\mathfrak{a}})$ *such that* $T(\mathfrak{a}) = T(\mathbf{a})$ *and, for each* $t \in T(\mathfrak{a})$ *and for each* $F_i^{\mathfrak{a}}$, $F_i^{\mathfrak{a}}(t) \in \Phi_i(\mathbf{a})$.

Let $\mathscr{P}$ be an empirical phenomenon in the physical sense, and let $Z_{\mathscr{P}} = \{\mathfrak{a}_p : p \in \mathscr{P}\}$, $\mathbf{Z}_{\mathscr{P}} = \{\mathbf{a}_p : p \in \mathscr{P}\}$ be two abstract representations of $\mathscr{P}$, the former consisting of numerical elementary systems and the latter of operational elementary systems. The two representations will be said to be *affined* if and only if for each $p \in \mathscr{P}$, $\mathfrak{a}_p \in Id(\mathbf{a}_p)$.

Call any set of the form $Z_{\mathscr{P}}$ a *numerical representation of* $\mathscr{P}$. In turn, sets of the form $\mathbf{Z}_{\mathscr{P}}$ will be referred to as *operational representations of* $\mathscr{P}$. The following is trivially valid.

ASSERTION 5.3.2. *A numerical representation* $Z_{\mathscr{P}}$ *of* $\mathscr{P}$ *is state-determined only if there is an operational representation* $\mathbf{Z}_{\mathscr{P}}$ *of* $\mathscr{P}$ *which is*: (a) *affined to* $Z_{\mathscr{P}}$, *and* (b) *operationally state-determined.*

A bit less obvious is

ASSERTION 5.3.3. *An operational representation* $\mathbf{Z}_{\mathscr{P}}$ *of* $\mathscr{P}$ *is operationally state-determined only if there is a numerical representation* $Z_{\mathscr{P}}$ *of* $\mathscr{P}$ *which is*: (a) *affined* to $\mathbf{Z}_{\mathscr{P}}$, *and* (b) *state-determined.*

PROOF. In order to prove the assertion one has to exploit the fact that when $\mathscr{P}$ is an empirical phenomenon it is at most a countable set. This allows to prove that there exist $\mathbf{Z}_{\mathscr{P}}$, $Z_{\mathscr{P}}$ and a one-to-one mapping f from $St(\mathbf{Z}_{\mathscr{P}})$ onto $St(Z_{\mathscr{P}})$ such that for each s in $St(Z_{\mathscr{P}})$, $f(s)$ is an idealization of s (i.e. if $s = (\Delta_1, \ldots, \Delta_n)$ and $f(s) = (x_1, \ldots, x_n)$, then $x_i \in \Delta_i$ for all $i = 1, \ldots, n$). Clearly $St(\mathbf{Z}_{\mathscr{P}})$ stands for the set of all states of all systems in $\mathbf{Z}_{\mathscr{P}}$. The representation $Z_{\mathscr{P}}$ is easily seen to have properties (a) and (b) in Assertion 5.3.3.

The proof has been presented in an outline but the parts omitted can easily be complemented.

The two assertions stated above should make it clear enough in what sense the ordinary (numerical) account of regularities may be claimed to be quivalent to the operational one.

5.4. THE CONCEPT OF TRUTH AS RELATED TO OPERATIONAL STRUCTURES

To have a deeper insight into what the theoretical consequences of the decision to represent empirical phenomena by operational systems instead of numerical ones are, let us now in turn examine how the approach proposed affects the classical conception of truth. Clearly, by the classical conception of truth we shall mean Tarski's explication of that conception.

The concept of an idealization of an operational structure will again play the key role in our analyses. Consider any operational structure of the form $\mathbf{a}_p^\tau$, where p is a phenomenon in the scope $\mathscr{K}$ of $\Theta_{\mathscr{K}}^\tau$. In view of the assumptions made at the very beginning of this chapter, each $\mathfrak{a}$ in $Id(\mathbf{a}_p^\tau)$ provides an incomplete interpretation for $\mathscr{L}_\Theta$ in the sense that it is a reduct of a partial interpretation for that language. Let us denote elements of $Id(\mathbf{a}_p^\tau)$ as $\mathfrak{a}^F$ instead of $\mathfrak{a}$. The latter symbol will be reserved for partial interpretations of $\mathscr{L}_\Theta$.

Put

$$\tilde{c}(\mathbf{a}_p^\tau) = \{\mathfrak{a}\colon \mathfrak{a}^F \in Id(\mathbf{a}_p)\}.$$

Although the symbol $\tilde{c}(\mathbf{a}_p^\tau)$ is new, the class for which it stands is already well known to us. We have made use of classes of this sort several times in various comments and examples.

Let us in turn define:

$$c(\mathbf{a}_p^\tau) = \{\mathfrak{a}\colon \mathfrak{a}/u \in \tilde{c}(\mathfrak{a}_p^\tau), \text{ for some } u \in T(\mathfrak{a})\}.$$

It is immediately seen that the class $\tilde{c}(\mathbf{a}_p)$ is a subclass of $c(\mathbf{a}_p)$. The latter consists of all systems whose time restrictions are in the former.

This somewhat unexpected modification of the definition of the correspondence function c needs some justification. In order to see why this modification has been necessary consider the following.

ASSERTION 5.4.1. *$(\Theta, \tau, \mathscr{K})$ is semantically sound in $p \in \mathscr{K}$ if and only if there exists an interpretation $\mathfrak{a} \in c(\mathbf{a}_p)$ such that $\Theta \subseteq Val(\mathfrak{a})$ (all theorems of Θ are valid in $\mathfrak{a}$).*

PROOF. Assume that $(\Theta, \tau, \mathscr{K})$ is semantically sound in $p \in \mathscr{K}$ Then the union set $\Theta \cup E_p$ is semantically consistent. Let $\mathfrak{a}$ be a model of $\Theta \cup E_p$ and thus E_p in particular. For each $(F_i, t, \Delta) \in E_p$ we have $F_i^{\mathfrak{a}}(t) \in \Delta$, where $F_i^{\mathfrak{a}}$ is the denotation of F_i in $\mathfrak{a}$. Hence $F_i^{\mathfrak{a}}(t) \in \Phi_i^{\tau}(t)$, and thus $\mathfrak{a}/T(p)$ is an idealization of $\mathbf{a}_p^{\tau}$. In general, however, $T(p)$ need not coincide with $T(\mathfrak{a})$. Consequently, $\mathfrak{a}$ need not be an element of $c(\mathbf{a}_p^{\tau})$ but, anyway, it is an element of $c(\mathbf{a}_p^{\tau})$ incidentally, this is the very reason why we should exploit the function c rather than $\tilde{c}$.

In turn, assume that $\mathfrak{a} \in c(\mathbf{a}_p)$ and $\Theta \subseteq Val(\mathfrak{a})$. The first part of this assumption yields $E_p \subseteq Val(\mathfrak{a})$. We have then $\Theta \cup E_p \subseteq Val(\mathfrak{a})$ which, by Definition 5.2.2, implies that $(\Theta, \tau, \mathscr{K})$ is semantically sound in p, and concludes the proof.

In order to see more clearly why the semantic soundness of $(\Theta, \tau, \mathscr{K})$ should be linked with $c(\mathbf{a}_p^{\tau})$ rather than with $\tilde{c}(\mathbf{a}_p^{\tau})$ assume that Θ involves a theorem (say an existential one) which is valid only in those systems whose duration is sufficiently long. One may easily give an example of such a theorem. Observe now that it is customary to take it for granted that whenever Θ is applicable to p, it is also applicable to any time restriction of p. Indeed, if the theory Θ can be applied to predict the behaviour of p in the whole period $T(p)$, it certainly can be applied to predict the behaviour of p in any subinterval u of $T(p)$. We wanted the notion of soundness to provide an explication of the informal notion of applicability (in the sense in which the latter has been used in the comments stated above).

For the considerations which follow it is of no importance at all how operational representations of the phenomena in the scope $\mathscr{K}$ of Θ are defined. We need only to be sure that to each p in $\mathscr{K}$ there corresponds exactly one df-system $\mathbf{a}_p$, without presupposing whether $\mathbf{a}_p = \mathbf{a}_p^{\tau}$ or not. We need not go into details how the correspondence function c has been defined in particular whether the definition of c coincides with that of $\tilde{c}$ or not. Again, what will bother us is only that for each $\mathbf{a}_p$, $p \in \mathscr{K}$, the set $c(\mathbf{a}_p)$ is defined. In that way we shift to a purely semantic ground, but one has to keep in mind that whatever is going to be said has an obvious and immediate operational counterpart. It is enough to assume that the df-systems dealt with are of the form $\mathbf{a}_p^\tau$, where τ are testing procedures of Θ, to introduce again operational aspects into the discussion.

In contradistinction to the notion of soundness which is related to the theory as a whole, the notion of truth has been understood as related to singular sentences. These two notions being discussed, it is often taken for granted that a theory in the semantic sense is a theory whose all theorems are true. Note however that except for this very special case of $c(\mathbf{a}_p)$ being a unit set (i.e. the case when there is exactly one standard interpretation corresponding to the operational structure $\mathbf{a}_p$), we must treat all structures in $c(\mathbf{a}_p)$ as equally good candidates to the role of that very structure with respect to which the notion of truth is to be defined, clearly unless some special conventions have been adopted which help to make the "proper" choice. The case of $c(\mathbf{a}_p)$ being a unit set is very special indeed. The operational structure $\mathbf{a}_p$ has exactly one idealization $\mathfrak{a}/_F$ only if for each Φ_i in the characteristic of $\mathbf{a}_p$ and for each t in $T(p)$, $\Phi_i(t)$ has as its value an interval of the form $[x, x]$. As we have already noticed, operational structures of that form may be identified with numerical ones. In turn, in order for $\mathfrak{a}/_F$ to have only one extension, $\mathfrak{a}/_F$ must be a partial interpretation for $\mathscr{L}_\Theta$ by itself. This still does not guarantee that $c(\mathbf{a}_p)$ contains only one element unless the duration of $\mathbf{a}_p$ is unlimited, i.e. coincides with $(-\infty, +\infty)$.

If we decide to treat all systems in $c(\mathbf{a}_p)$ as indiscernible (equally capable of being treated as systems with respect to which the notion of truth is to be defined), then the only acceptable solution to the problem of how to define truth values of sentences of $\mathscr{L}_\Theta$ seems to be the following.

(TR) Given any operational system $\mathbf{a}_p, p \in \mathscr{K}$, and given any sentence α of $\mathscr{L}_\Theta$ we define α to be:

(t) *true* in $\mathbf{a}_p$ if and only if $c(\mathbf{a}_p) \neq \emptyset$ and α is valid in all interpretations I in $c(\mathbf{a}_p)$,

(f) *false* in $\mathbf{a}_p$ if and only if $c(\mathbf{a}_p) \neq \emptyset$ and α is valid in no interpretation I in $c(\mathbf{a}_p)$,

(i) *indeterminate* in $\mathbf{a}$ if and only if α is neither true not false in $\mathbf{a}_p$.

Note that to be false does not amount to not to be true, although, as one may easily verify, α is false in the sense of the definition proposed above if and only if $\neg \alpha$ is true. Note also that the truth of $\alpha \vee \beta$ does not imply that α is true or β is true, although the converse is valid, i.e. if any of the constituents of the disjunction $\alpha \vee \beta$ is true, then $\alpha \vee \beta$ is also true. The falsity of $\alpha \vee \beta$ implies the falsity of both α and β, but the falsity of $\alpha \wedge \beta$ does not suffice to claim that either α or β is false.

Yet, in spite of all these peculiarities, the conception of truth set up by Definition (TR) is a simple, unpretentious and natural modification of Tarski's idea. The indeterminateness of the truth value of some sentences caused for instance by vagueness of some of the terms they involve has always been regarded as one of the indispensable attributes of natural languages. In particular, Łukasiewicz's interest in many-valued logics was motivated by his belief that the dychotomy of truth and falsity underlying classical logic does not stand critical examination.

In connection with this last remark however it should be stressed that the indeterminateness in the sense defined above cannot be treated as the "third truth-value", at least it cannot be treated as such if we expect the truth value of compound sentences to be the function of the truth-values of their constituents. Note for instance that the disjunction of two indeterminate sentences can be indeterminate (when they are logically independent of each other) or true (if for instance they are of the form α and $\neg \alpha$) (For further analyses of the relations between truth values and indeterminateness cf. M. Przełęcki, 1969, 1977, P. Williams, 1974, K. Fine, 1975).

We arrive at a more complicated but still realistic conception of truth by assuming that all or some theorems of $\Theta_{\mathscr{K}}$ play the role of conventions (meaning postulates) determining the way in which the specific concepts they involve should be interpreted. By a *meaning postulate* we shall understand any sentence α of $\mathscr{L}_\Theta$ such that by a semantic convention valid in $\mathscr{L}_\Theta$ (i.e. being a semantic rule of that language) all or only some of the terms involved in α must be interpreted under any intended application of the theory so that α is true (K. Ajdukiewicz, 1958, R. Carnap, 1952, 1958a). This explanation is put in a loose, informal language, but still it will serve our purposes quite well. Actually it merely has to provide an intuitive justification for the next definition. In order to define the notion of a meaning postulate in a precise way it would be necessary to go into some preparatory and rather involved considerations, in particular it would be necessary to define in a precise way the notion of a semantic rule (convention).

DEFINITION 5.4.2. *Assume that* $\pi \subseteq \Theta$ *is the set of meaning postulates of* $\mathscr{L}_\Theta$. *Given an operational system* $\mathbf{a}_p$, $p \in \mathscr{K}$, *and given any sentence* α *of* $\mathscr{L}_\Theta$ *we define* α *to be*

(t) true in $\mathbf{a}_p$ under π *if and only if there exists a partial interpretation* $\mathfrak{a}$ *in* $c(\mathbf{a}_p)$ *such that* $\pi \subseteq Val(\mathfrak{a})$ *and* α *is valid in all such interpretations.*

(f) false in $\mathbf{a}_p$ under π *if and only if there exists a partial interpretation* $\mathfrak{a}$ *in* $c(\mathbf{a}_p)$ *such that* $\pi \subseteq Val(\mathfrak{a})$ *and* α *is valid in no such interpretation.*

(i) indeterminate in $\mathbf{a}_p$ under α *if and only if* α *is neither true nor false in* $\mathbf{a}_p$ *under* π.

It is immediately seen that we arrive at (TR) by letting $\pi = \varnothing$. One may also easily see that if α is true under the empty set of meaning postulates, then α is true under any set of meaning postulates and similarly if α is false under the empty set of meaning postulates, then α is false under any set π. The sentences indeterminate under postulates π are indeterminate under $\varnothing$.

One of the immediate and, in fact, counterintuitive consequences of the definition proposed is that the meaning postulates need not be true! Given an operational structure $\mathbf{a}_p$ it may happen that for no $a \in c(\mathbf{a}_p)$, $\pi \subseteq Val(a)$. Note that in such a case all sentences of $\mathscr{L}_\Theta$ become indeterminate in $\mathbf{a}_p$ under π, including mathematical identities! This certainly is a hardly tolerable consequence of Definition 5.4.2 which as a matter of fact, indicates that meaning postulates should not be entirely arbitrary sentences. We shall discuss the matter in more detail in the next section. For the time being let us compare once more the notion of truth and that of soundness.

Suppose that all theorems of Θ are considered to be meaning postulates for the specific terms they involve. In general, given any domain $p \in \mathscr{K}$ the set of interpretations $c(\mathbf{a}_p)$ comprises more than one structure, and in general only some of the interpretations I in $c(\mathbf{a}_p)$ are such that $\Theta \subseteq Val(I)$. In that way our decision to treat Θ as a set of meaning postulates distinguishes some interpretations among all comprised by $c(\mathbf{a}_p)$ and thus contributes to setting up the meanings of the specific terms of Θ. It perfectly makes sense to say that the meanings of the specific terms of a theory are determined among others by the theorems in which they are involved, provided that saying that we do not deny the role of other factors which may influence our way of understanding the specific terms of Θ. Were theorems of Θ the only source of the meaning for those terms, the theory Θ would be purely formal.

In connection with the last remark the reader may be anxious to know what exactly is to be meant by the "meaning" of a term. Indeed, one can scarcely come across any notion which is as often employed in philosophical discussion as "meaning" is, curiously enough quite frequently without taking a sufficent care to make it clear how it should be understood. To some extent we adhere to this blameworthy tradition. Whenever the word "meaning" appears in our comments it is applied in a loose informal manner. But there is no convincing evidence that the notion of meaning, being very helpful in all informal comments, is indispensable when the discussion of methodological issues is carried out in rigorous terms. In fact it seems to be of minor significance how and to what extent theorems, testing procedures, our decision concerning the scope of the theory, common sense and perhaps some further methodological, psychological and sociological factors contribute to determining the meaning of the terms of the theory. What counts is how all these factors should be interrelated among themselves in order for the theory to be good. Now if experimental results concerning a domain p in $\mathscr{K}$ and theorems of $\Theta^{\tau}_{\mathscr{K}}$ are in discrepancy there is no doubt that $\Theta^{\tau}_{\mathscr{K}}$ cannot be kept unchanged. It is just the case when $\Theta^{\tau}_{\mathscr{K}}$ is not sound.

To restore the soundness of a theory we have to modify some or even all of its constituents. We may find it necessary to redefine the content Θ of $\Theta^{\tau}_{\mathscr{K}}$. Perhaps we decide to change testing procedures τ. We may also revise our views on what should be counted as the scope $\mathscr{K}$ of the theory. Whatever modifications are carried into effect, the resulting theory $(\Theta', \tau', \mathscr{K}')$ should be sound.

Of a secondary importance seems to be also the concept of a meaning postulate. After all, when a sentence α is inconsistent with the results of experiment, it does not make any difference whether α is just a theorem or also a meaning postulate. In both the cases such an inconsistency indicates that the theory is not sound and some or even all of its consistuents should be modified. The argument fails when the set of meaning postulates π is selected so that its elements are not capable of empirical falsification, i.e. π is consistent with any set E of empirically decidable hypotheses, provided that E is consistent. Still the defence of the notion of a meaning postulate carried out along this line is not convincing, either. In order for a theory not to be sound with respect to a domain p the existence of a theorem inconsistent with E_p is not necessary. It is enough that $\Theta \cup E_p$ is not consistent, even if for each α in Θ, $\{\alpha\} \cup E_p$ is. Now if Θ is to be changed in order to restore the soundness of $\Theta^{\tau}_{\mathscr{K}}$, the meaning postulates (definitions including!) are as good candidates for modification as any of the remaining theorems. Thus, when dynamical aspects are taken into account the role and the status of a meaning postulate in the theory does not seem to be essentially different form those of any theorem whatsoever.

The fact that the discrepancy between the theory and experience need not involve the discrepancy between a particular theorem and experience is well known. It was a keystone of conventionalist account of science (H. Poincare, P. Duhem), and it motivated so-called *coherentional conception of truth* ('Die Wahrheit bildet ein System' H. Weyl, 1927). By deciding to consider soundness of a theory rather than truth of its theorems as a criterion for evaluating theories we accomodate certain elements of conventionalist and coherential account of science though, the very fact that we discern between soundness and truth shows that the approach we have adopted can be classified neither as conventionalistic nor as coherentialistic.

5.5. TRUTH BY CONVENTION

The notion of a meaning postulate is often treated as unseparable from that of a theoretical term. Since there are no empirical procedures by means of which elementary sentences involving theoretical terms can be decided, it is natural to assume that the meaning of theoretical terms is set up by means of postulates (which may happen to acquire the form of definitions). At the same time all the remaining terms are often treated as defined "operationally", "by ostensive definitions" or just as "observational" terms.

The way in which we have defined empirical terms does not allow to treat them as observational. The idea of an ostensive definition originated from an attempt (not very promising, I am afraid) to give an account of how some simple quantitative terms can be furnished with meaning. It is rather obvious that empirical terms as we want to understand them need not be ostensively definable. They can be operationally definable but they certainly have some theoretical meaning in the sense we tried to render in the preceding section.

The reader should be warned then against identifying the notion of an observational with that of an empirical term. Still in what follows the empirical terms of $\Theta_{\mathcal{X}}^{\tau}$ will be treated as if they were observational. We shall ignore the fact that empirical terms may very well be theoretically laden and in particular some theorems stated in empirical terms can be notoriously empirically undecidable. (For an example consider the theorem of particle mechanics which says that the position is a twice differentiable function with respect to time).

Assuming that the meaning of empirical terms is fixed we shall try to describe how the theorems of Θ determine the meaning of theoretical terms. This is exactly the problem which was discussed by R. Carnap in his papers on the notion of analyticity (cf. especially (1958a)). We have to start then with presenting his view on the matter.

In our account of Carnap's view we shall ignore certain divergences between the nature of $\mathscr{L}_\Theta$ and the languages discussed by R. Carnap; he restricted himself to elementary languages, and he contrasted theoretical terms just with observational ones. The divergences mentioned are of minor significance for the heart of the matter. Following R. Carnap let us assume that Θ is finitely axiomatizable. Let us assume that $P_1, \ldots, P_r$ are all axioms of Θ which involve theoretical terms. Denote those terms as $G_1, \ldots, G_m$. Let

(1) $$P(G_1, \ldots, G_m)$$

stand for the conjunction of $P_1, \ldots, P_r$.

Under the assumptions we made, saying that the meaning of the terms G_i is determined by theorems of Θ amounts to stating that it is determined by the postulates $P_1, \ldots, P_r$, or equivalently by conjunction (1). Clearly the role of the postulates $P_1, \ldots, P_n$ consists in the fact that given an interpretation for empirical symbols (say a system in $Id(\mathbf{a}_p)$), they determine the class of all interpretations for the theoretical terms under which they are true. Note however that this class may happen to be empty! For an illustration of the last remark consider the theory *LH* defined in Chapter III. The existence of such an interpretation for the symbol *D* under which Axioms D1–D7 (cf. Definition 3.2.2) become true depends on the interpretation acquired by the terms we decided to regard as empirical.

When selecting interpretations for theoretical terms we must pay attention to how empirical terms have been interpreted, which clearly mirrors the fact that $P_1, \ldots, P_n$ are not merely postulates for $G_1, \ldots, G_m$ but also some factual theorems. In particular, some of the consequences of $P_1, \ldots, P_n$ may happen to be empirically decidable. Following Carnap, let us call those hypotheses whose truth-value depends only on semantic conventions *analytic* and all the remaining ones *factual* (or *synthetical*). We shall also employ the abbreviation *A-true* for *analytically true* and *F-true* for *factually true*. Now the conclusion at which we arrive examining the restriction imposed by postulates $P_1, \ldots, P_n$ on interpretations for theoretical terms can be worded as follows. The very fact that $P_1, \ldots, P_n$ are meaning postulates does not guarantee that they are A-true.

Is it then possible to separate analytic and factual hypotheses and in particular to divide the theorems of Θ into A-true and F-true? Carnap's solution to the problem is perhaps the simplest one to be proposed in the case when the set of meaning postulates is assumed to be finite, although it is not the only possible one (for details cf. M. Przełęcki and R. Wójcicki, 1969).

Replace in (1) $G_1, \ldots, G_m$ by variable symbols $x_1, \ldots, x_m$ respectively and denote the existential closure of the resulting formula as $\sum(P)$, i.e. $\sum(P)$ stands for

(2) $$\exists x_1 \ldots \exists x_m P(x_1, \ldots, x_m).$$

The sentence (2) is occasionally called *Ramsey sentence*.

The conjunction

(3) $$\sum(P) \wedge (\sum(P) \rightarrow P)$$

is immediately seen to be logically equivalent to P. Since $\sum(P)$ does not involve theoretical terms, its truth value may not depend on conventions concerning those terms. Then $\sum(P)$ is a factual hypothesis.

On the other hand all empirical terms which appear in the second constituent of (3), i.e. in the sentence $\sum(P) \rightarrow P$, do not appear in that sentence in an essential way, in the sense that the following assertion is valid:

ASSERTION 5.5.1. *For each standard interpretation I for $\mathscr{L}_\Theta$ there exists a standard interpretation I' for that language such that*:

(i) *the restrictions of I and I' to empirical terms coincide, and*

(ii) $\sum(P) \rightarrow P$ *is valid under the interpretation I'.*

PROOF. The proof of the assertion is trivial. Indeed, select any standard interpretation I for $\mathscr{L}_\Theta$. If $\sum(P)$ is not valid under that interpretation, then clearly $\sum(P) \to P$ is. In turn, if $\sum(P)$ is valid under I, then there are objects $G_1^I, \ldots, G_m^I$ which satisfy the propositional formula $\mathrm{P}(x_1, \ldots, x_m)$, $G_i^{I'}$ being assigned to x_i respectively. The interpretation I' such that $I'(F_i) = I(F_i)$, $i = 1, \ldots, n$, and $I'(G_j) = G_j^{I'}$, $j = 1, \ldots, m$, is the interpretation under which both the formulas $\sum(P)$ and P are valid. Hence $\sum(P) \to \mathrm{P}$ is also valid. This concludes the proof.

Thus it is enough to take as a postulate for theoretical terms the sentence $\sum(P) \to P$ to be sure that there exists an interpretation under which $\sum(P) \to P$ is valid, regardless of how the empirical terms are interpreted. Note also that Assertion 5.5.1 yields the following:

COROLLARY 5.5.2. *For each sentence* α, *if* α *does not involve any theoretical terms, then*

$$\sum(P) \to P \vdash \alpha$$

if and only if $\vdash \alpha$.

PROOF. If $\vdash \alpha$, then clearly $\sum(P) \to P \vdash \alpha$. On the other hand if $\nvdash \alpha$, then for some I, $\alpha \notin Val(I)$. Take any I' for which Conditions (i) and (ii) of Assertion 5.5.1. are satisfied. Clearly $\alpha \notin Val(I')$ by (i) and at the same time $\sum(P) \to P \in Val(I')$. Thus $\sum(P) \to P \nvdash \alpha$ which concludes the proof.

Given any set π of sentences which involve both theoretical and empirical terms, we call π *conservative in the semantic sense* with respect to empirical terms if and only if each interpretation for empirical terms can be extended to a standard interpretation I for $\mathscr{L}_\Theta$ such that $\pi \subseteq Val(I)$. More precisely: for each standard interpretation I there exists an interpretation I' such that the two interpretations restricted to empirical terms coincide and $\pi \subseteq Val(I')$.

In turn, if all sentences which both are derivable from π and do not involve theoretical terms are mathematically provable, the set π is said to be *conservative in the syntactical sense* with respect to empirical terms.

Applying the terminology we have introduced we can state Theorems 5.5.1 and 5.5.2 saying that $\sum(P) \to P$ is both semantically and syntactically conservative with respect to empirical terms. It is worth noticing that semantic conservativeness implies the syntactical one but, in general, the converse is not true (for a more detailed discussion of the issue cf. M. Przełęcki and R. Wójcicki, 1971).

The way in which the problem of analyticity has been solved by R. Carnap as well as some of his comments on it allow to state some general requirements which should be satisfied by any solution to that problem.

Let λ_A and λ_F be two disjoint sets of sentences of L_Θ (call them *A-component* and *F-component* of Θ, respectively) which satisfy the following conditions:

(λ1) The union $\lambda_A \cup \lambda_F$ is semantically (syntactically) equivalent to Θ.

(λ2) λ_A is semantically (syntactically) conservative with respect to empirical terms.

(λ3) λ_F does not involve theoretical terms.

Conditions (λ1) and (λ2) have been stated ambiguously. They admit both semantical (stronger) and syntactical (weaker) interpretation, but in our concise and very general comments on the issue we need not pay much attention to the consequences of selecting either of these interpretations.

If the sets λ_A and λ_F exist, and Carnap's considerations show that they do exist, at least in the case when Θ is finitely axiomatizable, one may define a theorem $\alpha \in \Theta$ to be analytic (A-true) if and only if $\lambda_A \models \alpha$ (or, if the syntactic interpretation of (λ1)–(λ3) is accepted, if and only if $\lambda_A \vdash \alpha$).

Curiously enough it turns out that Θ may possess many non--equivalent analytical components and thus, unless (λ1)–(λ3) are suplemented by some additional conditions, there is no unique solution to the problem of analyticity.

I do not want to get involved into technicalities to which the issue discussed leads, some of them being quite sozhisticated. The reader who may happen to become interested in the problem of analyticity is recommended to consult rather rich literature on it (For bibliographical references cf. e.g. Wójcicki, 1975).

5.6. CONFIRMATION PROCEDURES

There is no doubt that by taking dispersive procedures into consideration we shall make a step towards a more realistic account of relationships between phenomena and their theories. Data collected by means of dispersion free procedures cover, as a rule, only a modest part of empirical evidence available, even in the case of phenomena which can be examined with the help of dispersion free procedures, not to mention those which can be subjected to statistical examination only.

As we have already emphasized, the very fact that the experiment is not dispersion free does not suffice to question its usefulness in science. It becomes useless only if the data gathered by successive repetitions of the experiment escape any reasonable statistical processing, i.e. when there is no standarized and commonly accepted (by the experts) way to summarize them in the form of a statistical assertion. The definition stated below should provide a more rigorous account of conditions we want the considered testing procedures to satisfy.

DEFINITION 5.6.1. *A function* $\pi: H_\pi \to [0, 1]$, *where* $H_\pi \subseteq H$, *will be said to be a* confirmation function *if and only if the following two conditions are satisfied*:

(π1) *There exists a finitely additive probability measure P defined on the Lindenbaum-Tarski algebra* $(H/\sim, \cup, \cap, -)$ *such that* $P(|\alpha|) = \pi(\alpha)$, for all α in H_π.

(π2) *If* $\alpha_1, \ldots, \alpha_k$ *are in* H_π, β *is in H and there exists* x_0 *such that for each probability measure P defined on the algebra* $(H/\sim, \cup, \cap, -)$ $P(|\beta|) = x_0$ *whenever*

$$P(|\alpha_1|) = \pi(\alpha_1), \ldots, P(|\alpha_n|) = \pi(\alpha_n),$$

then β *is in* H_π *and* $\pi(\beta) = x_0$.

We obviously have

COROLLARY 5.6.2. *If* π *is a confirmation function such that for each* α *in* H_π, *either* $\pi(\alpha) = 1$ *or* $\pi(\alpha) = 0$, *then* π *is a verification function.*

In what follows the verification function will often be referred to as a *dispersion-free confirmation function.*

Definition 5.6.1. is self-explaining. Still let us briefly discuss conditions ($\pi 1$) and ($\pi 2$).

From the logical point of view any two sentences which are deductively equivalent convey the same information. An equivalence class of the form $|\alpha|$ may be viewed then as the "content" of α or as the proposition represented by the sentence α. Consequently, the algebra $(H/_\sim, \cup, \cap, -)$ may be called an *algebra of propositions.* The fact that the values which π assigns to experimental-like sentences may be interpreted as probabilities of propositions makes it possible to view π itself as a probability measure.

The following assertion is trivially valid.

ASSERTION 5.6.3. *The set of the following conditions is equivalent to* ($\pi 1$): *For all* α, β *in* H_π:

($\pi 3$) $0 \leqslant \pi(\alpha) \leqslant 1$.

($\pi 4$) $\pi(\alpha) = 1$, *whenever* $\vdash \alpha$.

($\pi 5$) *If* $\alpha \vdash \neg \beta$ *and* $\alpha \vee \beta$ *is in* H_π, *then* $\pi(\alpha \vee \beta) = \pi(\alpha) + \pi(\beta)$.

($\pi 6$) *If* $\alpha \vdash \beta$, *then* $\pi(\alpha) \leqslant \pi(\beta)$.

The proof of the assertion is easy and we shall omit it.

Observe that ($\pi 3$)$-$($\pi 5$) are analogons of the postulates for finitely additive probability measure. Condition ($\pi 6$) tells how the derivability relation $\vdash$ and confirmation function π are related to each other.

($\pi 2$) provides a characteristic of the domain H_π of π. For instance the following conditions:

($\pi 7$) *If* $\alpha \in H_\pi$, *then* $\neg \alpha \in H_\pi$.

(π8) *If* $\alpha_1, \ldots, \alpha_n$ *are in* H_π, $\pi(\alpha_1) = \ldots = \pi(\alpha_n) = 1$, β *is in* H, *and* $\alpha_1, \ldots, \alpha_n \vdash \beta$, *then* $\beta \in H_\pi$.

cannot be proved without the help of (π2). As a matter of fact (π1) does not presuppose anything about H_π but non-emptiness of that set.

In order to have a deeper insight into the nature of the conditions (π1), (π2), denote as πH the set of all sentences of the form $\pi(\alpha) = x$, with $\alpha \in H$ and $x \in [0, 1]$. Clearly, the sentences in πH are not sentences of $\mathscr{L}_\Theta$, they are rather elements of a certain metalanguage of the language $\mathscr{L}_\Theta$; we shall refer to them as *π-sentences.*

Observe that each probability space

(1) $(H, H/_\sim, P)$,

where $H/_\sim$ is the Lindenbaum-Tarski algebra defined on H, and P is a probability measure defined on $H/_\sim$, may be viewed as a model-structure for πH. Indeed, given any π-sentence '$\pi(\alpha) = x$' one may define it to be valid in $(H, H/_\sim, P)$ if and only if $P(|\alpha|) = x$. Consequently, the set πH considered together with all probability spaces of the form defined above can be treated as a language. In particular one may define all semantic notions for πH in the standard way. For instance, given a set χ of π-sentences we shall say that χ is semantically consistent if and only if there is a probability space of the form (1) in which all π-sentences in χ are valid. The set χ will be said to *entail* a π-sentence σ if and only if σ is valid in each probability space of the form (1) in which all π-sentences in χ are valid.

In order to emphasize that the semantic notions of the sort mentioned above are defined relative to structures which in an essential way involve a Lindenbaum-Tarski algebra we shall refer to those notions as *LT-notions.* Thus we shall speak about *LT-consistency*, *LT-entailment*, etc..

Denote by *Diag*(π) the subset of πH defined by the following condition:

(*) A π-sentence of the form '$\pi(\alpha) = x$' is in *Diag*(π) if and only if $\alpha \in H_\pi$ and $\pi(\alpha) = x$.

The set $Diag(\pi)$ will be referred to as a *diagram of* π.

In order for the definitions given above to make sense we need not assume that π is a confirmation function; it is enough to postulate that $\pi: H_\pi \to [0, 1]$, where $H_\pi \subseteq H$.

As an almost immediate corollary from Definition 5.6.1 we have

THEOREM 5.6.4. *A function* $\pi: H_\pi \to [0, 1]$, *where* $H_\pi \subseteq H$, *is a confirmation function if and only if*

(i) $Diag(\pi)$ *is LT-consistent.*
(ii) *If* $Diag(\pi)$ *LT-entails a* π*-sentence* σ *then* $\sigma \in Diag(\pi)$.

One may easily see that (π1) is equivalent to (i) and (π2) is equivalent to (ii).

Theorem 5.6.4 is a "dispersive" counterpart of Theorem 5.1.9, which was proved when nondispersive testing procedures were discussed. As a matter of fact 5.6.4 is not only a counterpart of 5.1.9, but a generalization of the latter. When π is dispersion free, Theorems 5.6.4 and 5.1.9 become equivalent.

Some additional comments on Theorem 5.6.4 will be given in the next section.

NOTE. The assumption that the results of experiment can be uttered in the form of sentences $\tau(\alpha) = x$ gives rise to the following doubt. Quite often the experimenter is not able to establish the exact probability of the tested hypothesis though he is able to estimate it. Should not then the results of experiments be viewed as sentences of the form

$$\pi(\alpha) \in \Delta,$$

where Δ is a subinterval of $[0, 1]$, rather than in the form $\tau(\alpha) = x$, the only one we decided to take into account?

The reason why we have decided to concentrate our attention on the ideal situation, thus resigning from the very beginning from a more realistic account should be obvious. A discussion of the ideal case indirectly covers the situations in which the experimenter is not able to establish the exact value of τ in the same way as empirical theories stated in terms of precise values of the specific terms they involve cover the situations in which experimental results provide only approximate, or still worse only statistical, account of the state of affairs under examination.

5.7. PROBABILISTIC MODELS

An important difference betweenthe dispersive case we are discussing and the dispersion-free one with which we dealt in the previous chapter consists in the fact that the empirical evidence concerning a given domain p cannot, in general, be expressed without a symbol for the function τ_p. Note that if τ_p is dispersion free this can easily be done. The sentence α may be treated as a translation of the sentence $\tau(\alpha) = 1$, and the sentence $\neg \alpha$ may be treated as a translation of $\tau(\alpha) = 0$. Incidentally, in that way E_p turns out to be the translation of $Diag(\tau_p)$ *in* $\mathscr{L}_\Theta$.

Since in the dispersive case the symbol for τ_p cannot be eliminated we must define correspondence rules which will allow us to relate the content Θ of $\Theta_{\mathscr{X}}$ to the empirical evidence available, i.e. to $Diag(\tau_p)$. The simplest way in which this can be done is to introduce the symbol τ to $\mathscr{L}_\Theta$ and to assume that given any sentence $\alpha \in \mathscr{L}_\Theta$, the sentence α and the sentence $\tau(\alpha) = 1$ of the new language are translations for each other.

The solution to the problem raised we have in mind is suggested by the considerations of the previous section. Let us present it in a more rigorous way.

Denote as $\tau\mathscr{L}_\Theta$ the set of sentences of the form $\tau(\alpha) = x$, where $\alpha \in \mathscr{L}_\Theta$ and $x \in [0, 1]$. In what follow the sentences of that sort will be referred to as *τ-sentences.* By an *LT-model structure for* $\tau\mathscr{L}_\Theta$ we shall understand a finitely-additive probability space of the form

$$(\mathscr{L}_\Theta, \mathscr{L}_\Theta/_\sim, P).$$

The notion of validity of a τ-sentence in an LT-model structure for $\tau\mathscr{L}_\Theta$ should be defined in the same way as that of validity of a π-sentence in an LT-model structure for πT was.

DEFINITION 5.7.1. *A sentence* '$\tau(\alpha) = x$' *of* $\tau\mathscr{L}_\Theta$ *is* LT-valid *in a LT-model structure* $(L_\Theta, \mathscr{L}_\Theta/_\sim, P)$ *for* $\tau\mathscr{L}_\Theta$ *if and only if* $P(|\alpha|) = x$.

Also the notion of LT-entailment and that of LT-consistency of a set of τ-sentences are to be defined in full analogy to their πT-counterparts.

One should realize that the approach proposed above is too restrictive. To see this assume that τ is non-dispersive (i.e. takes only the ultimate values 0 and 1). In that particular case the language $\tau\mathscr{L}_\Theta$ becomes translatable onto $\mathscr{L}_\Theta$ in the way which we already described: α serves as the translation of $\tau(\alpha) = 1$ and vice versa.

With the help of Theorem 1.11.6 one may prove

ASSERTION 5.7.2. *Let $X \subseteq \mathscr{L}_\Theta$, $\alpha \in \mathscr{L}_\Theta$. Denote as X^τ the set of all τ-sentences of the form $\tau(\beta) = 1$, where $\beta \in X$.* Then

$$X \vdash \alpha \text{ if and only if } X^\tau \models \tau(\alpha) = 1.$$

PROOF. Assume that $X \vdash \alpha$, then $X \cup \neg\alpha$ is deductively inconsistent. Thus for some $\beta_1, \ldots, \beta_n$ in X, the conjunction $\beta_1 \wedge \ldots \wedge \beta_n \wedge \neg\alpha$ is deductively inconsistent. This implies that in each probability space of the form $(L_\Theta, L_\Theta/_\sim, P)$

$$(1) \qquad P(|\beta_1 \wedge \ldots \wedge \beta_n \wedge \neg\alpha|) = 0.$$

Suppose now that $P(|\beta|) = 1$ for each β in X, then clearly by (1) we have

$$P(|\neg\alpha|) = 0,$$

or equivalently

$$P(|\alpha|) = 1.$$

Thus $X^\tau \models \tau(\alpha) = 1$, which concludes the first part of the proof.

Assume in turn $X \nvdash \alpha$, and thus $X \cup \{\neg\alpha\}$ is consistent. Then, by Theorem 1.11.6 there exists a homomorphism V from the Lindenbaum-Tarski algebra onto the two-valued Boolean algebra such that $V(X) \subseteq 1$ and $V(\neg\alpha) = 1$, or equivalently $V(\alpha) = 0$. Note now that $(\mathscr{L}_\Theta, \mathscr{L}_\Theta/_\sim, V)$ is a probability space in which all τ-sentences of X^τ are valid but $\tau(\alpha) = 1$ is not. Hence $X^\tau \not\models \tau(\alpha) = 1$ which concludes the proof.

It turns out then that under the proposed interpretations for $\tau\mathscr{L}_\Theta$, semantic properties of $\tau\mathscr{L}_\Theta$ do not correspond to semantic properties of $\mathscr{L}_\Theta$ as one might have expected, but rather to syntactical ones. In order to define "genuinely" semantic relations between the sentences of $\mathscr{L}_\Theta$ (the laws of $\Theta_\mathscr{X}$ in particular) and empirical evidence stated in the form of τ-sentences we must adopt a radically different approach from that we have presented. The main idea of the semantics for probabilistic hypotheses we are going to discuss is borrowed from J. Łoś (1963).

Given any set $\mathcal{J} \subseteq Int(\mathcal{L}_\Theta)$, denote as $Mod_{\mathcal{J}}(X)$ the class of all standard interpretations in $\mathcal{J}$ in which X is valid. We let then

$$Mod_{\mathcal{J}}(X) = \{I \in \mathcal{J} : X \subseteq Val(I)\}.$$

Definition 5.7.3. *Let* $(\mathcal{J}, \mathfrak{J}, P)$, *where* $\mathcal{J} \subseteq Int(\mathcal{L}_\Theta)$, *be a probability space. We shall say that* $(\mathcal{J}, \mathfrak{J}, P)$ *is a* probabilistic interpretation (model) for $\mathcal{L}_\Theta$ *if and only if for each set of sentences* X, $Mod_{\mathcal{J}}(X) \in \mathfrak{J}$.

It is immediately seen that the concept of a probabilistic interpretation is more general than that of an interpretation and thus all semantic notions relevant for $\mathcal{L}_\Theta$ can be redefined with the help of it. In particular the notion of validity can be defined as follows.

Definition 5.7.4. *A set of sentences* X *of* $\mathcal{L}_\Theta$ *will be said to be* valid in a probabilistic interpretation $(\mathcal{J}, \mathfrak{J}, P)$ of $\mathcal{L}_\Theta$ *if and only if* $P(Mod_{\mathcal{J}}(X)) = 1$.

We certainly have the following

Assertion 5.7.5. *A set of sentences* X *of* $\mathcal{L}$ *is semantically consistent if and only if there exists a probabilistic interpretation* $(\mathcal{J}, \mathfrak{J}, P)$ *of* $\mathcal{L}$ *such that* $P(Mod_{\mathcal{J}}(X)) = 1$, *i.e.* X *is valid under* $(\mathcal{J}, \mathfrak{J}, P)$.

Proof. Assume first that X is semantically consistent. Then there is a realization I of X. Put $\mathcal{J} = \{I\}$, $\mathfrak{J} = \{\emptyset, \{I\}\}$, $P(\emptyset) = 0$, $P(I) = 1$. Obviously $(\mathcal{J}, \mathfrak{J}, P)$ is a probabilistic interpretation, and obviously $P(Mod_{\mathcal{J}}(X)) = P(I) = 1$. On the other hand, if for some $(J, \mathfrak{J}, P)$, $P(Mod_{\mathcal{J}}(X)) = 1$, then $Mod_{\mathcal{J}}(X) \neq \emptyset$ (since $P(\emptyset) = 0$) and thus X is semantically consistent.

As an immediate and obvious corollary from Assertion 5.7.5. we have

ASSERTION 5.7.6. *Let* $X \subseteq \mathscr{L}$ *and* $\alpha \in \mathscr{L}$. *Then* $X \models \alpha$ *if and only if for each probabilistic interpretation* $(\mathscr{I}, \mathfrak{J}, P)$ *if* X *is valid under this interpretation, then also* α *is valid.*

ASSERTION 5.7.7. *A set of sentences* X *of* $\mathscr{L}$ *is semantically consistent if and only if there exists a probabilistic interpretation* $(\mathscr{I}, \mathfrak{J}, P)$ *of* $\mathscr{L}$ *such that* $P(Mod_{\mathscr{I}}(X)) = 1$, *i.e.* X *is valid under* $(\mathscr{I}, \mathfrak{J}, P)$.

PROOF. Assume first that X is semantically consistent. Then there is a realization I of X. Put $\mathscr{I} = I$, $\mathfrak{J} = \{\emptyset, \{I\}\}$, $P(\emptyset) = 0$, $P(I) = 1$. Obviously, $(\mathscr{I}, \mathfrak{J}, P)$ is a probabilistic interpretation, and obviously $P(Mod_{\mathscr{I}}(X)) = P(I) = 1$. On the other hand if for some $(\mathscr{I}, \mathfrak{J}, P)$, $P(Mod_{\mathscr{I}}(X)) = 1$, then $Mod_{\mathscr{I}}(X) \neq \emptyset$ (since $P(\emptyset) = 0$) and thus X is semantically consistent.

As an immediate and obvious corollary from Assertion 5.7.7 we have

ASSERTION 5.7.8. *Let* $X \subseteq \mathscr{L}$ *and* $\alpha \in \mathscr{L}$. *Then* $X \models \alpha$ *if and only if for each probabilistic interpretation* $(\mathscr{I}, \mathfrak{J}, P)$ *if* X *is valid under this interpretation, then also* α *is valid.*

It is immediately seen that each probability space of the form $(\mathscr{I}, \mathfrak{J}, P)$ may serve as a model structure for $\tau\mathscr{L}_\Theta$. The definition of the notion of validity of a τ-sentence in such a model structure results from Definition 5.7.1 by an obvious transformation. Still we gave up the idea of the language τL_Θ and we decided to speak about probabilistic models for $\mathscr{L}_\Theta$ rather than about models for $\tau\mathscr{L}_\Theta$. The decision was motivated by the fact that in $\tau\mathscr{L}_\Theta$ one cannot form statements which refer to probabilities of sets of sentences (except for sets of sentences which may be axiomatized by a single formula).

NOTE. Let $X \subseteq \mathscr{L}_\Theta$. It is worth noticing that in general the fact that X^τ is valid in an Ł-model structure (X^τ being defined on the set of all τ-sentences of the form $\tau(\alpha) = 1$, with $\alpha \in X$) need not imply that X is valid in that structure. To put it in other words, the fact that $P(Mod_{\mathscr{I}}(\alpha)) = 1$ for all $\alpha \in X$, need not imply that $P(Mod_{\mathscr{I}}(X)) = 1$.

We are in a position now to state the definition of soundness of $(\Theta, \tau, \mathscr{K})$.

DEFINITION 5.7.9.

A. *The theory* $(\Theta, \tau, \mathscr{K})$ *is* deductively sound with respect to $p \in \mathscr{K}$ *if and only if for some* P,

$$(\mathscr{L}_\Theta, \mathscr{L}_\Theta/\sim, P)$$

is a probability space such that the following two conditions hold true:

(i) $P(|\alpha|) = 1$, *for all* α *in* Θ,
(ii) $P(|\alpha|) = \tau(\alpha)$, *for all* α *in* H_p.

B. *The theory* $(\Theta, \tau, \mathscr{K})$ *is* semantically sound in a domain $p \in \mathscr{K}$ *if and only if there exists a probabilistic interpretation* $(\mathscr{I}, \mathfrak{J}, P)$ *for* $\mathscr{L}_\Theta$ *such that the following two conditions are satisfied*:

(j) $P(Mod_{\mathscr{I}}(\Theta)) = 1$,
(jj) *for each* α *in* H_p, $P(Mod_{\mathscr{I}}(\alpha)) = \tau(\alpha)$.

Note that the "black and white" picture of relations between the theory and empirical evidence characteristic of "traditional" semantics need not be kept unchanged. The probabilistic approach allows us to order the theories in accordance with how well they conform to the results of the experiment.

DEFINITION 5.7.10.

A. *We shall say that* ϱ *is the* degree of deductive reliability of $(\Theta, \tau, \mathscr{K})$ in an application $p \in \mathscr{K}$ *if and only if* ϱ *is the greatest number in* $[0, 1]$ *such that for some* P,

$$(\mathscr{L}_\Theta, \mathscr{L}_\Theta/\sim, P)$$

is a finitely additive probability space in which, for all $\varepsilon > 0$, *the following two conditions are satisfied*:

(d1) $P(|\alpha|) \geqslant \varrho - \varepsilon$, *for all* α *in* Θ,
(d2) $P(|\alpha|) = \tau(\alpha) \pm 1 - \varrho + \varepsilon$, *for all* α *in* H_p.

B. ϱ *is the* degree of semantic reliability of $(\Theta, \tau, \mathscr{K})$ in $p \in \mathscr{K}$ *if and only if* ϱ *is the greatest number in* $[0, 1]$ *such that for some probabilistic interpretation* $(\mathscr{I}, \mathfrak{J}, P)$ *for* $\mathscr{L}_\Theta$, *and for all* $\varepsilon > 0$, *the following two conditions are satisfied*:

(s1) $P(Mod_{\mathscr{I}}(\Theta)) \geqslant \varrho - \varepsilon$,
(s2) $P(Mod_{\mathscr{I}}(\alpha)) = \tau(\alpha) \pm 1 - \varrho + \varepsilon$, *for all* α *in* H_p.

We have the following

ASSERTION 5.7.11. *The semantic degree of reliability of* $(\Theta, \tau, \mathscr{K})$ *in* $p \in \mathscr{K}$ *is not greater than the deductive one.*

PROOF. Assume that $(\mathscr{I}, \mathfrak{J}, P)$ is a probabilistic interpretation for $\mathscr{L}_\Theta$ for which conditions (s1), (s2) are satisfied. Define the function P' on $\mathscr{L}_\Theta/_\sim$ by the condition

$$P'(|\alpha|) = P(Mod_{\mathscr{I}}(\alpha)).$$

We have to prove that P' is well defined and it is a probability measure in $\mathscr{L}_\Theta/_\sim$. Assume first that $\alpha \sim \beta$. Then clearly $Mod_{\mathscr{I}}(\alpha) = Mod_{\mathscr{I}}(\beta)$ and hence $P'(|\alpha|) = P'(|\beta|)$. This proves that $P'(|\alpha|)$ does not depend on a particular selection of the sentence which determines the equivalence class $|\alpha|$, and hence P' is well defined.

We have to prove that: (a) $P'(|\alpha \vee \neg \alpha|) = 1$ and (b) If $|\alpha| \cap |\beta| = |\alpha \wedge \beta| = |\alpha \wedge \neg \alpha|$, then $P'(|\alpha \vee \beta|) = P'(|\alpha|)+P'(|\beta|)$. (a) is trivially valid since $Mod_{\mathscr{I}}(\alpha \vee \neg \alpha) = \mathscr{I}$. Suppose now that $|\alpha \wedge \beta| = |\alpha \wedge \neg \alpha|$. Then $\alpha \vdash \neg \beta$ and consequently $Mod_{\mathscr{I}}(\alpha) \cup Mod_{\mathscr{I}}(\beta) = \emptyset$. On the other hand, $Mod_{\mathscr{I}}(\alpha \vee \beta) = Mod_{\mathscr{I}}(\alpha) \vee Mod_{\mathscr{I}}(\beta)$, and since $Mod_{\mathscr{I}}(\alpha) \cap Mod_{\mathscr{I}}(\beta) = \emptyset$, then $P(Mod_{\mathscr{I}}(\alpha \vee \beta)) = P(Mod_{\mathscr{I}}(\alpha))+P(Mod_{\mathscr{I}}(\beta))$ which immediately yields (b) as it is needed, concluding the proof.

Definitions 5.7.9 and 5.7.10 immediately yield the following:

COROLLARY 5.7.12. $(\Theta, \tau, \mathscr{K})$ *is deductively (semantically) sound with respect to* $p \in \mathscr{K}$ *only if the deductive (semantic) degree of reliability of* $(\Theta, \tau, \mathscr{K})$ *in* p *equals* 1.

The question under what conditions deductive soundness of $(\Theta, \tau, \mathscr{K})$ with respect to p implies semantic soundness has got no easy answer. Even much more modest problem of what the conditions under which there exists such a probabilistic model $(\mathscr{I}, \mathfrak{J}, P)$ that for each α in H_p $P(Mod_{\mathscr{I}}(\alpha)) = \tau(\alpha)$ are, does not seem to have any ready solution.

Some problems closely related to those mentioned now have been discussed by J. Łoś (1963). Here we shall not try to pursue the issues further.

5.8. DISPERSIVE OPERATIONAL STRUCTURES

The fact that testing procedures need not be dispersion free makes the idea of representing phenomena in the scope $\mathscr{K}$ of $\Theta_{\mathscr{K}}$ by operational systems inadequate. In that situation we shall rediscuss the notion of an operational system and propose a certain improved version of it.

One certainly may question the very expediency of any considerations of that sort and argue that nothing forces us to give up the convenient although unrealistic assumption that for each p in $\mathscr{K}$ there exists an elementary structure $\mathfrak{a}_p$ providing an ideal representation for p. Obviously the structure $\mathfrak{a}_p$ is not defined by the empirical evidence available, i.e. by the τ-sentences which form the set $Diag(\tau_p)$, still $Diag(\tau_p)$ serves as a sort of an incomplete probabilistic "definition" (or rather diagram) of $\mathfrak{a}_p$.

Suppose however that we insist on representing domains of $\Theta_{\mathscr{K}}$ by structures which are completely and uniquely defined by the empirical facts we are able to establish. Certainly, if the system which represents p is assumed to be defined by τ, each modification of τ, for instance any improvement of testing procedures, yields a modification (improvement) of the system representing p. Of considerable importance becomes then the problem of examining what the "limit" of the sequence of all such modifications of systems modelling p might be, and in particular whether the limit system may be viewed as an ideal structure, i.e. a numerical elementary structure.

The way in which that new sort of operational representations for empirical phenomena should be defined is rather obvious. Denote as Ω the set of all functions φ such that for some $B \subseteq \boldsymbol{B}(Re)$,

(1) $\varphi : B \to [0, 1]$.

Let u be an interval of real numbers and let $\Gamma_1, \ldots, \Gamma_n$ be functions of the form

(2) $\Gamma_i : u \to \Omega$.

The structure

(3) $(u, \Gamma_1, \ldots, \Gamma_n)$

will be referred to as an *elementary dispersive operational system*, *d-system* for short.

In view of an obvious analogy between d-systems and elementary systems one may easily redefine the notions applicable to the latter ones so as to make them applicable to the former. For instance the state of the system (3) at $t \in u$ should be defined as the sequence

$$(\varphi_1, \ldots, \varphi_n)$$

with

$$\varphi_i = \Gamma_i(t).$$

DEFINITION 5.8.1. *A d-system* $\delta_p^\tau = (u, \Gamma_1, \ldots, \Gamma_n)$ *will be said to be a* τ-model *of the domain* p *of* $\Theta_{\mathscr{K}}$ *if and only if the following two conditions are satisfied*:

(φ1) $u = T(p)$.

(φ2) *For each* Γ_i *and for each* t *in* u, *if* $\Gamma_i(t) = \varphi$, *then* $\varphi(\Delta)$ *is defined if and only if* (F_i, t, Δ) *is in* H_p *and moreover* $\varphi(\Delta) = \tau_p(F_i, t, \Delta)$.

As we have already mentioned, the evolution of science brings constant improvements of testing procedures

$$\tau, \tau', \tau'', \ldots$$

and consequently constant modifications of d-structures

$$\delta_p^\tau, \delta_p^{\tau'}, \delta_p^{\tau''}, \ldots$$

which serve as operational dispersive models for p. Let us try to make it more precise in what sense, given two sets of testing procedures τ^i, τ^j, one may claim that those from one of the sets, say τ^j, are better than the procedures from the other.

The problem raised involves at least two different aspects. When evaluating τ^i and τ^j one may be primarily interested in soundness of the procedures the two sets involve, and consequently one may judge τ^i and τ^j according to the "degree of their soundness". But as I shall argue it may happen very well that both the procedures in τ^i and those in τ^j are sound, though the confirmation functions τ_p^i, τ_p^j they determine do not coincide. Thus apart from soundness some other criteria of goodness of testing procedures may be of some importance. As such a criterion we shall define and discuss the "degree of dispersiveness".

Let us start with some comments on the notion of soundness of testing procedures. From an intuitive point of view a testing procedure is sound when empirical data gathered with the help of it are adequate. Clearly, in order to judge the adequacy of data we must have an independent way of verifying them. Thus soundness of some experimental procedures can be appreciated only under the assumption that some others are sound.

An alternative way to evaluate adequacy of the data collected and thus soundness of the procedures applied when gathering them consists in verifying whether the data fit the theory, i.e. whether the theory assumed to involve the pertinent procedures is sound.

In practice, the two ways of evaluating soundness of testing procedures are often applied simultaneusly. We reject or modify those tests which yield results contradicting well established laws, and, when such a possibility exists, we verify one testing procedure against another.

The informal remarks made above, although clear enough, I believe, when referred to some dispersion free procedures, may require some additional comments in the dispersive case being of special interest for us at present.

The question which may be raised is how one may verify correctness of data collected by means of τ^i with the help of data collected by means of τ^j when the procedures involved in both τ^i and τ^j are dispersive.

We shall restrict ourselves to brief and informal comments on the matter. Let us assume first that the procedures τ^j are dispersion free and furthemore each hypothesis decidable with the help of τ^i is decidable with the help of τ^j. Let X^ξ be the set of all hypotheses τ^i-decidable and such that $\tau^i(\alpha) = \xi$ for each α in X^ξ. Draw at random a finite subset $X_0^\xi \subseteq X^\xi$ and verify with the help of τ^j which of the hypotheses in X_0 are valid (more exactly τ^j-valid). In order for τ^i to be sound with respect to τ^j the experiments concerning sets of the form X_0^ξ should result in establishing that approximately 100 ξ per cent of all sentences in such sets are valid. Clearly, any attempts to implement the procedure described must lead to many both technical and theoretical problems we have deliberately ignored.

In turn, assume that both τ^i and τ^j are dispersive. In that case in order for τ^i to be sound relative to τ^j (and thus τ^j to be sound relative to τ^i!) it is necessary that one could assign truth values to the hypotheses which are both τ^i- and τ^j- decidable in such a way that both τ^i and τ^j are sound with respect to this (arbitrary!) assignment in exactly the same sense in which τ^i has been sound with respect to (the assignment defined by) τ^j in the previous case.

We have concentrated ourselves on the criteria of soundness, but it should be easily seen that soundness can be graduated. Indeed, if soundness of τ^i and τ^j is evaluated in the context of the theory $\Theta_{\mathscr{K}}$ whose laws we accept as unquestionable, then τ^j may be claimed to be at least as sound as τ^i if the degree of reliability of $(\Theta, \tau^j, \mathscr{K})$ is in all applications of that theory not less than that of $(\Theta, \tau^i, \mathscr{K})$.

It also makes sense to say that τ^j is sounder relative to τ^k than τ^i (the way in which such a locution is to be understood is rather obvious), and obviously if τ^k is assumed to be sound, one may omit mentioning τ^k and simply say that τ^j is *sounder* than τ^i.

Let us make however a rather unrealistic assumption that at each stage of the development of the theory the procedures τ, $\tau', \tau'', \ldots$ it involves are sound. In that case, given two sets τ^i, τ^j we certainly would prefer the one which allows us to decide empirical hypotheses in a "more conclusive" way. If the probability of α evaluated with the help of τ^i is say 0.62 while $\tau^j(\alpha) = 0.93$, both the practical and theoretical value of τ^j-information is certainly greater than that of τ^i-information (let me stress once again that both τ^i and τ^j are assumed to be sound!). The surer the gathered data are, the better, both when a practical action is planned and when the data are intended to serve as a base for further theorizing.

Now the notion of degree of conclusiveness, or as we shall

prefer to say *degree of dispersiveness* of testing procedures, involved in the comments above, can be explicated as follows. Given two sets of confirmation procedures τ^i, τ^j, both being applicable to p, we shall say that τ^j is less dispersive than τ^i in the application p whenever

$$H_p^i \subseteq H_p^j,$$

where H_p^i and H_p^j are the sets of τ^i-decidable and τ^j-decidable hypotheses, respectively, and furthermore for each $\alpha \in H_p^i$

$$\min(\tau^i(\alpha), \tau^i(\neg\alpha)) \geqslant \min(\tau^j(\alpha), \tau^j(\neg\alpha)).$$

One may view the sequence $\tau, \tau', \tau'', \ldots$ as consisting of less and less dispersive procedures tending to become non-dispersive.

In spite of a very loose informal character of the discussion we have carried out above the following conclusion seems to be quite well motivated. The dispersive nature of testing procedures does not make it senseless to assume that given any phenomenon p in the scope of Θ the progress of science and experimental techniques will eventually make it possible to define in a unique way an elementary system $\mathfrak{a}_p$ being the proper abstract representation of p.

Clearly just like in the case of approximate procedures this assumption is a sort of wishful thinking. But again as in the case of approximate procedures it provides us with a simple and useful idealization of the actual state of affairs. Though the concept of a d-model can be useful in examining formal properties of empirical theories we still may pretend that a satisfactory representation of empirical phenomena is provided by systems of "traditional" sort, i.e. numerical systems.

The situation will become radically different when we assume that dispersiveness of testing procedures is not caused by technical deficiency of the instruments applied or imperfectness of experimental schema or other removable factors but is due to the intrinsic nature of the phenomena examined. In that case the idea of an empirical system should be revised once again and in a considerably deeper way than it was done when the two operational counterparts of numerical systems were defined.

We shall briefly discuss the matter in the next chapter.

5.9. EVOLUTION OF EMPIRICAL THEORIES

In order to view a theory as a dynamical system rather than as a stationary one, one should simply assume that all its components (language, content, testing procedures, scope and whichever else one decides to distinguish) depend on time. Thus, for instance, under this approach the theory in the stationary sense $(\Theta, \tau, \mathcal{K})$ we have dealt with should be viewed as the state of a theory in the dynamical sense $(\Theta(t), \tau(t), \mathcal{K}(t))$ at a fixed instant of time t_0, i.e.

$$(\Theta, \tau, \mathcal{K}) = (\Theta(t_0), \tau(t_0), \mathcal{K}(t_0)).$$

One should realize that a genuinely dynamical account of empirical theories cannot be formed in the way we described. It is simply not enough to take a new variable "time" into considerations in order to arrive at a theory of regularities characteristic of the evolution of sciences. Were such a theory possible it should enable us to answer questions concerning the behaviour of dynamical systems of the form

$$(\Theta(t), \tau(t), \mathcal{K}(t)),$$

in particular to predict at least in a statistical manner their future states on the base of data concerning the states already reached.

May anybody seriously expect methodology of empirical theories to answer questions like the following:

(Q1) What are the modifications which the conceptual apparatus of Θ will undergo in the period of time $[t_0, t_1]$?

(Q2) Call α to be unrefutable if $\alpha \in \Theta(t)$ implies that $\alpha \in \Theta(t')$ for all $t' \geqslant t$. Which of the laws of $\Theta(t_0)$ are unrefutable?

(Q3) Let F be a specific variable of Θ. What is the greatest accuracy in measuring F which will be achieved in the course of the evolution of T?

Obviously not! The methodology is not a sort of a super-science which enables us to predict the future history of science. Were such predictions possible, methodologists would be able to solve certain problems specific for a given theory even before the solution is known to scientists.

It is fashionable to speak about dynamics of empirical theories, one ought to discern, however, between inquiry into dynamics of empirical theories which should result in discovering deterministic regularities and studies of intertheory relations in particular comparative studies of different stages in the development of the same (in the dynamical sense) theory. Although the latter unavoidably involves some dynamical aspects they need not result in establishing any deterministic laws whose predictive power resembles in the slightest way the predictive power of deterministic laws one may find in natural sciences.

Comparative studies carried out in methodology of empirical theories are of specific sort. One of the main functions of the methodology is to discover, to analyze, and to process the criteria which scientists apply, often in the intuitive way and sometimes even unwillingly, when evaluating their theories. Methodologists are then primarily interested in establishing conditions under which some theories deserve to be claimed to be better than others. In particular, all heuristic models of the evolution of empirical theories constructed and discussed by methodologists involve the assumption that scientists act rationally and thus if there are many ways in which a given theory might be changed, the best option is the one which will eventually be selected. The evolution of science is then notoriously viewed as falling under a sort of Principle of Constant Improvements, whose predictive value is almost null but explanatory role considerable.

Let me emphasize once more that soundness is not the only requirement, and perhaps is not even the most important one a good theory should satisfy. In evaluating theories instrumental criteria seem to be at least as important as semantic and operational ones. Each theory may be viewed as a device to solve some practical as well as theoretical problems. It is natural then to assume that given two theories Θ, Θ', both being applicable to the same class $\mathscr{K}$ of empirical phenomena, one of them, say Θ', is better (more useful) from the instrumental standpoint than the other if each problem which can be defined and solved in Θ may also be defined and solved in Θ'. The definition should by no means be treated as definite, it merely indicates the line along which our further discussion will proceed.

Let us begin with some remarks concerning the concept of a problem. First of all assume that we are interested only in those problems whose solution can always be expressed in the form of a sentence or a set of sentences. A problem is said to be closed, or have a closed class of answers, if the class of possible solutions to it is known. Among closed problems there are, for instance, all so-called whether-questions, i.e. questions to which a proper answer is "yes" or "no" e.g. "Does the number 3 divide 6587?". "Is the distance between the Earth and the Sun less than one billion miles?", "Does the parameter F affects behaviour of the system α?". The answers "yes" and "no" can be replaced by the pair of corresponding sentences, for instance "The number 3 divides 6587" and "The number 3 does not divide 6587", and obviously, such a pair constitutes the class of possible solutions to the problem.

Clearly, closed problems are not restricted to whether-questions. The question "What day is it today?" does not fall under this category (it is so-called which-question), nevertheless it expresses a certain closed problem. The class of possible solutions to it consists of the sentences "Today is Monday", "Today is Tuesday", ..., "Today is Sunday".

Quite often open problems, i.e. those which are not closed, concern causes or mechanisms of phenomena. Asking about a cause of some event E (e.g. What caused a sudden stop of the engine?), we may happen to be lucky enough to foreknow the full list of factors which can be responsible for the event. Clearly, in such a situation solving the problem reduces to discovering those factors from the list which bring about the event in question. Then, of course, the problem belongs to the class of closed systems.

As a rule, however, questions about a cause of an event are open, because we are unable to foretell all possible factors which account for it. Of open character are the questions concerning the way of preventing or discovering certain events (for instance those concerning methods of diagnosing an illness in its early stage). No doubt, without much difficulty the reader can quote a number of questions to which not only solutions but also possible solutions are not known.

In spite of an unquestionable significance of open questions, when comparing empirical theories we should concentrate upon closed questions since only under such a restriction it becomes possible to state definite criteria for deciding which of the theories compared is the best one from the instrumental standpoint. The restriction is substantia only seemingly. Observe that if α is an answer to an open question P, then P and the closed question "Is it true that α?" are equivalent in the sense that deciding one of them is equivalent to deciding the other and vice versa. Thus comparison of theories with respect to their usefulness for deciding closed questions comprises indirectly comparison of their usefulness for deciding problems of quite arbitrary type. Division of problems into open and closed intersects the division into hypothetical and actual. The former ones concern hypothetical situations, therefore in order to decide them one has first to enlist premises on the base of which a solution to them is to be sought for. Let us however confine ourselves to actual problems. Since they concern real situations, solving such a problem P on the grounds of a theory Θ requires collecting such empirical data $E_1, \ldots, E_n$ which together with certain theorems $T_1, \ldots, T_n$ of the theory allow to prove a sentence H being an answer to the question expressing the problem. In the case when the question is closed, the answer H should, of course, be an element of the class of all possible solutions to this question. If suitable empirical data and theorems of the theory allow to show that some of the possible solutions to the problem P are false, i.e. they allow to prove negation of those solutions, we shall say that the theory Θ enables partial solution to the problem.

Partial as well as complete solutions to problems are subject to evaluation with respect to their adequacy. An adequate solution need not be true; we are often satisfied with solutions which deviate from the actual state of affairs provided that an error made is not too big. Notice that if a theory Θ makes it possible to find "approximately adequate" solution to the problem P, it makes it possible to find a fully adequate (thus true) solution to the problem P after the latter has been modified in an appropriate way. Let us illustrate the point with a suitable example.

Suppose that the problem we want to decide consists in establishing the value which a quantity F will have at time t. Suppose further that the solution given takes the form of the following sentence

(1) $$F(t) = x_0 \pm \Delta.$$

As sentence (1) does not tell what the exact value of the quantity F is, the solution is partial. Assume additionally that it is false, i.e. the real value of F is outside the interval $x_0 \pm \Delta$. However, the error may happen to be negligibly small and from the practical point of view insignificant. Thus, as a matter of fact, although false, such a solution is quite satisfactory. However, it is clear that the formulation "What value will the quantity F take at time t?" can be replaced by a more specified question. For instance we may want the value of F at time t to be determined with some accuracy ε. Possible solutions to the problem so formulated must be of the form

(2) $$F(t) = x \pm \varepsilon.$$

In particular, the sentence

(3) $$F(t) = x_0 \pm \varepsilon$$

which, provided that $\varepsilon \geqslant \Delta$, follows from (1), may turn out to be not only complete but also true solution to the problem in its new wording.

It is obvious that if Θ' is an enlargement of Θ and both theories are sound, each problem which can be adequately solved on the ground of Θ can also be solved on the ground of Θ'. It is worth noticing that even if Θ' is a later stage of Θ (i.e. Θ' has resulted from a modification of Θ) the languages of the two theories may be considerably different. Passing from Θ to Θ' we could have enlarged the number of specific notions and replaced some of them with new ones. Also mathematical apparatus of Θ could have been modified. In such a situation comparison of the theories Θ and Θ' by comparing their contents need not lead to any interesting conclusions whatsoever. Much more can be learnt about the relations between the two theories by asking whether the problems decidable in Θ (all of them or perhaps belonging to a certain class, e.g. those which are relevant for a practical applications of the theory) preserve their decidability in Θ'.

Observe that if the theory Θ' allows to decide all problems which are decidable in Θ, that by no means implies that Θ' is an enlargement of Θ. The problem which is decidable in both Θ and Θ' can be decidable in them in a significantly different way. In particular, if the notions we operate with on the grounds of Θ differ from those of Θ', in order to solve a given problem with the use of theorems of the theory Θ it may be necessary to collect data quite different from those needed in the case of solving the same problem with the help of theorems of Θ'. It may then turn out that if the data needed when we operate with the theory Θ are easily available, the data necessary when the theory Θ' is being used are much more difficult to collect. The differences may be those of costs one must bear to carry out suitable experiments, or of time needed in order to complete them.

Since we have included cost of collecting empirical data necessary for solving a problem among the factors which should be taken into account when choosing between two theories, let us still make the following quite obvious remark. One of the stages of solving a problem consists in inferring a solution from theorems and collected empirical data. Clearly, kinds of theorems in the scope of the theory, their formulation as well as the kind of mathematical apparatus underlying fundamentals of the theory are decisive for the degree of difficulty one is faced with when trying to solve a problem under investigation. In this respect the difference between two theories that can occur may be not less important than those discussed before. It does matter whether a given problem can be solved with the help of an easy algorithm or whether complicated, non-typical, i.e. requiring special additional researches, mathematical methods are necessary.

The discussion we have presented gives an insight into the most important factors which we take into account when comparing theories and finally making up our mind to consider one of them as better or at least better in some respects than the other. Searching for improved theories we are looking for such which allow to solve a broader class of problems than a theory we actually have at our disposal. Obviously that is equivalent to searching for a theory with a broader scope on one hand and for theories grasping stronger regularities on the other. We are trying to find theories which in the case when no adequate solution is available, offer the most adequate solution attainable to a problem examined. We are searching for theories which lead to solutions laden with as small error as possible, and finally we are searching for theories which allow us in a relatively easy and inexpensive way to find a solution to the problems we wish to investigate. Unfortunately, very often those aims are irreconciliable. That is why apart from general theories concerning a very broad class of systems, special theories grasping pecularities of certain special cases are often formulated. Beside the theories which guarantee a high degree of precision but at the same time are quite complicated, theories based on simplifying assumptions are often used. The latter ones often consitute an earlier stage of development of a given theory, i.e. they are in a sense outdated theories, and a constant interest of them is due to certain special merits which are not inherited by their more perfect counterparts.

VI. APPENDIX

6.1. COMPLEMENTARY TESTS

The discussion to be carried out in this concluding part of the book will primarly be motivated by some problems characteristic of modern physics, notably of quantum mechanics. Our main task will be to reconsider once more the concept of an empirical system. Still, before we do that let us examine what consequences may be brought about as the result of dropping out an assumption which thus far we have tacitly been accepting in our considerations concerning testing procedures.

Call any two experimental hypotheses α, β *codecidable by testing procedures* τ if both α and β are decidable by means of procedures in τ and furthermore testing α does not prevent testing β and vice versa. Although the matter has not been discussed, actually we took if for granted that whenever any two hypotheses are decidable they are codecidable. The reader who is familiar with discussions concerning fundamentals of quantum mechanics knows that by the very nature of empirical phenomena it may happen that codecidability of certain hypotheses is impossible. Any couple of so-called complementary quantum-mechanical quantities (say, position and momentum) furnish us with suitable examples. If either of these two quantities is measured with a high degree of precision, the other can be measured only in an imprecise way. The interdependence of the available precisions of measurements is goverened by an appropriate physical law.

One may invent suitable "classical" examples of tests which cannot be carried out simultaneously and eventually result in non-codecidability of certain decidable hypotheses. What makes however the classical and quantum-mechanical cases different is that while in the former the lack of codecidability is caused by the nature of testing procedures (for instance the lack of codecidability of α, β may be yielded by the fact that they both refer to the same object which must be distroyed in order to complete the test), in the latter the particular selection of testing procedures does not affect the situation. The hypotheses which are not codecidable in the quantum-mechanical sense are such regardless of any particular selection of testing procedures. It is the physical nature of phenomena to which the hypotheses refer rather than our measurement capabilities which is decisive in this respect.

The quantum-mechanical intuitions will underlie our further considerations, at least in the sense that we shall define the notion of codecidability so that it will be independent of any particular selection of testing procedures.

Following the commonly accepted terminology, two codecidable hypotheses α, β will be referred to as *commeasurable*, in symbolic notation $\alpha \,ó\, \beta$. The following two assumptions have an obvious motivation.

(ó1) *The commeasurability relation is reflexive and symmetric.*

(ó2) *If* $\alpha \,ó\, \beta$, $\alpha \sim \alpha'$, $\beta \sim \beta'$ *then* $\alpha' \,ó\, \beta'$.

The commeasurability relation will be assumed to be defined on the set H of all experimental-like sentences of $\mathscr{L}_\Theta$. Given any subset $H' \subseteq H$, we shall say that H' is *codecidable* if and only if for all $\alpha, \beta \in H'$, $\alpha \,ó\, \beta$. Note that one may as well take as a primitive concept the concept of codecidability and define $\alpha \,ó\, \beta$ to hold true when for some codecidable set $H' \subseteq H$, $\alpha, \beta \in H'$.

The next and the last condition we shall impose on the two notions being discussed is:

(ó3) *If* $H' \subseteq H$ *is a codecidable set, so is* $H' \cup \{\neg \alpha : \alpha \in H'\} \cup \{\alpha \vee \beta : \alpha, \beta \in H'\}$.

By Kuratowski–Zorn's lemma one may easily prove that for each codecidable subset H' of H there exists a maximal codecidable subset $\bar{H}$ of H such that $H' \subseteq \bar{H}$. Indeed, form any chain

$$H'_1 \subseteq H'_2 \subseteq \ldots$$

of commeasurable subsets of H such that $H'_1 = H'$ and consider the union $\bigcup_i H'_i$. The union is easily seen to be a codecidable subset of H again. Thus each chain of the considered sort has an upper bound in the class of all codecidable subsets of H containing H'. The class must then contain a maximal element.

Consider the quotient set $H/_\sim$, where $\sim$ denotes, as usual, the deductive equivalence of sentences. Define:

(ȸ) $|\alpha| \text{ȸ} |\beta|$ *if and only if* $\alpha \text{ȸ} \beta$.

($-$) $-|\alpha| = |\neg \alpha|$.

(1) $1 = |\alpha \vee \neg \alpha|$.

Demand also the following condition to be satisfied

($\cup$) *If* $\alpha \text{ȸ} \beta$, *then* $|\alpha| \cup |\beta| = |\alpha \vee \beta|$.

Condition ($\cup$) defines then the operation $\cup$ only partially; the result of applying the operation $\cup$ to $|\alpha|$ and $|\beta|$ is not defined when not $\alpha \text{ȸ} \beta$.

The following theorem is of some interest:

THEOREM 6.1.1. *The structure* $(H/_\sim, \text{ȸ}, \cup, -, 1)$ *with* ȸ, $\cup$, $-$, and 1 *defined as in the conditions stated above is a partial Boolean algebra in the sense of S. Kochen and E. P. Specker* (1965), *i.e. for all* a, b, c *in* $H/_\sim$ *the following conditions are satisfied*:

(i) *The relation* ȸ *is reflexive and symmetric* (*i.e.* $a \text{ȸ} a$ *and* $a \text{ȸ} b$ *implies* $b \text{ȸ} a$).

(ii) $a \text{ȸ} 1$,

(iii) *If* $a \text{ȸ} b$, *then* $a \text{ ȸ} -b$.

(iv) $a \cup b$ *is defined if and only if* $a \text{ȸ} b$ (*i.e.* ȸ *is the domain of* $\cup$)

(v) *If* a, b, c *bear the relation* ȸ *to each other, then* $a \cup b \text{ȸ} c$

(vi) *If* a, b, c *bear the relation* ȸ *to each other, then the set of all elements of* $H/_\sim$ *which result from* a, b, c *by successive applications of operations* $\cup$ *and* $-$ (*the set of Boolean polynomials from* a, b, c) *is a Boolean algebra with the unit* 1.

PROOF. The reflexivity and symmetricity of ȸ defined on $H/_\sim$ is yielded immediately by (ȸ) and the reflexivity and symmetricity of ȸ defined on H.

Note that given any sentence α, $\alpha \text{ȸ} \alpha \vee \neg \alpha$ in virtue of (ȸ3). On the other hand, since $\alpha \vee \neg \alpha \sim \beta \vee \neg \beta$, then by (ȸ2)

$\alpha \phi \beta \vee \neg \beta$. This yields $|\alpha| \phi |\alpha \vee \neg \alpha|$ or equivalently $|\alpha| \phi 1$, which establishes (ii).

Conditions (iii) and (v) say that the relation ϕ defined on $H/_\sim$ is closed under the operations $\cup$ and $-$. But this again is an immediate consequence of an analogous property of ϕ defined on H, cf. (ϕ3).

Condition (iv) reflects the fact that ($\cup$) defines $|\alpha| \cup |\beta|$ only when $\alpha \phi \beta$.

Since conditions (iii) and (v) are satisfied, all Boolean polynomials from a, b, c are defined. Now, the fact that whenever $-|\alpha|$ and $|\alpha| \cup |\beta|$ are defined they are defined in exactly the same way as in the Tarski-Lindenbaum algebra (cf. Assertion 5.1.2) guarantees that (vi) is satisfied, i.e. the algebra generated from a, b, c by the operations $\cup$ and $-$ is a Boolean algebra.

The very fact that some hypotheses $\alpha, \beta \in H$ are not commeasurable does not rule out the possibility to include both α and β into the set H_p of hypotheses concerning p and decidable by means of τ. Still we ought to remember that any attempt to evaluate the theory $\Theta_{\mathscr{X}}$ in an application p should be carried out with respect to hypotheses in an arbitrarily selected intersection $\bar{H} \cap H_p$, $\bar{H}$ being a maximal codecidable subset of H, not in the whole set H_p. Clearly, if H is codecidable, the only maximal codecidable subset of H is H itself. If H is not codecidable we shall refer to procedures τ as *complementary*, though this terminology may be misleading! We have already pointed out that the failure of H to be codecidable need not have anything to do with the testing procedures applied.

The following definition has an obvious justification.

DEFINITION 6.1.2. *Let τ be a set of complementary procedures, and for each p, let τ_p be a verification function. The theory $(\Theta, \tau, \mathscr{K})$* is deductively (semantically) sound in $p \in \mathscr{K}$ *if and only if for each maximal codecidable subset $\bar{H} \subseteq H$, the union $\Theta \cup (E_p \cap \bar{H})$ is* deductively (semantically) consistent.

Definition 5.2.2 is an immediate corollary from the last one, provided that H is assumed to be codecidable

6.2. PHYSICAL SYSTEMS

A notion of the physical system we are going to define will be strongly related to the current discussion on fundamentals of quantum mechanics. The definition is quoted after Mączyński (1968). It is a variant of a similar (though not equivalent) definition proposed by some other authors (cf. G. Mackey (1963), S. Gudder (1967)).

To begin with let us generalize the notation (F, t, Δ) as follows. Let S be the set of all possible states of the system to which the variable F refers and let $\xi \in S$, then (F, ξ, Δ) will be assumed to read: the variable F measured at the state ξ has a value in Δ.

In what follows the symbol $\mathcal{O}$ will be applied to denote a fixed set of empirical variables, say all or some selected variables of a given theory Θ.

Given any couples $(F, \Delta), (F', \Delta')$ in $\mathcal{O} \times \boldsymbol{B}(Re)$ define

$$(F, \Delta) \sim (F', \Delta)$$

to hold true if and only if for each ξ in S

$$P(F, \xi, \Delta) = P(F', \xi, \Delta').$$

It is immediately seen that $\sim$ is an equivalence relation on $\mathcal{O} \times \boldsymbol{B}(Re)$. We shall denote the equivalence class determined by (F, Δ) as $|(F, \Delta)|$, which as a matter of fact, is the standard notation. Clearly we have

$$|(F, \Delta)| = \{(F', \Delta) : (F', \Delta') \sim (F, \Delta)\}.$$

The set of all equivalence classes of the form $|(F, \Delta)|$, i.e. the quotient set $\mathcal{O} \times \boldsymbol{B}(Re)/_{\sim}$, will be denoted as $\mathfrak{E}$.

The definition we are going to state will involve the notion of an *$\mathfrak{E}$-measure*. It will be convenient however to postpone defining it till some later part of this section. For the time being let us merely explain that an *$\mathfrak{E}$-measure* is a function μ: $\boldsymbol{B}(Re) \to \mathfrak{E}$ of a certain specific sort.

DEFINITION 6.2.1. *Let $\mathcal{O}$, and S be non-empty sets and let*

$$P: \mathcal{O} \times S \times \boldsymbol{B}(Re) \to [0, 1].$$

The triple $(\mathcal{O}, S, P)$ *will be said to be an* abstract physical system *if and only if the following axioms are satisfied*:

Axiom 1. *For all* $\mathbf{E}$ *in* $\mathcal{O}$ *and all* ξ *in* S,

$$P(E, \xi, Re) = 1$$

Axiom 2. *If* $\mathbf{E}$ *is in* $\mathcal{O}, \xi$ *in* S *and* $\Delta_1, \Delta_2, \dots$ *is an infinite sequence of Borel sets such that* $\Delta_i \cap \Delta_j = \varnothing$, *for all* $i \neq j$, *then*

$$P(E, \xi, \Delta_1 \cup \Delta_2 \cup \dots) = \sum_{i=1} P(E, \xi, \Delta_i).$$

Axiom 3. *If* $P(E, \xi, \Delta) = P(E', \xi, \Delta)$ *for all* ξ *in* S *and all* Δ *in* $\mathbf{B}(Re)$, *then* $\mathbf{E} = \varepsilon'$.

Axiom 4. *If* $P(E, \xi, \Delta) = P(E, \xi', \Delta)$ *for all* $\mathbf{E}$ *in* $\mathcal{O}$ *and all* Δ *in* $\mathbf{B}(Re)$, *then* $\xi = \xi'$.

Axiom 5. *If* $E_1, E_2, \dots$ *are in* $\mathcal{O}, \Delta_1, \Delta_2, \dots$ *are in* $\mathbf{B}(Re)$ *and for all* $i \neq j$ *and* ξ *in* S, $P(E_i, \xi, \Delta_i) + P(E_j, \xi, \Delta_j) \leqslant 1$, *then there exist* E *in* $\mathcal{O}$ *and* Δ *in* $\mathbf{B}(Re)$ *such that for all* ξ *in* S,

$$P(E, \xi, \Delta) = \sum_{i=1} P(E_i, \xi, \Delta_i).$$

Axiom 6. *Let* $\mu: \mathbf{B}(Re) \to \mathfrak{E}$ *be an* $\mathfrak{E}$*-measure. Then there is* E *in* $\mathcal{O}$ *such that* $\mu(\Delta) = (|E, \Delta|)$, *for all* Δ *in* $\mathbf{B}(Re)$.

Before we comment on the definition stated above let us introduce some auxiliary notions. Define

$$|(E_1, \Delta_1)| \leqslant |(E_2, \Delta_2)|$$

to hold true if and only if for each ξ in S

$$P(E_1, \xi, \Delta_1) \leqslant P(E_2, \xi, \Delta_2),$$

and put

$$|(E_1, \Delta_1)|^{\perp} = |(A, R-E)|.$$

The following theorem is known.

THEOREM 6.2.2. *If* $(\mathcal{O}, S, P)$ *is a physical system*, $\mathfrak{E} = (\mathcal{O} \times S)/\tilde{-}$, *then the structure* $(\mathfrak{E}, \leqslant, {}^{\perp})$ *is a partially ordered set with orthocomplement, i.e. the following conditions are satisfied by all* a, b, c *in* $\mathfrak{E}$:

(PO1) $a \leqslant a$.
(PO2) *If* $a \leqslant b$ *and* $b \leqslant c$, *then* $a \leqslant c$.
(PO3) $(a^{\perp})^{\perp} = a$.
(PO4) *If* $a_1 \leqslant a_2$, *then* $a_2^{\perp} \leqslant a_1^{\perp}$.
(PO5) *If* $a_1, a_2, \ldots$ *are in* $\mathfrak{E}$ *and* $a_i \leqslant a_j^{\perp}$ *for all* $i \neq j$, *then there is the least element* a *in* P *such that* $a_i \leqslant a$, *for all* $i = 1, 2, \ldots$. *i.e.* $a = a_1 \cup a_2 \cup \ldots$.
(PO6) $a \cup a^{\perp} = b \cup b^{\perp}$ *for all* a, b *in* $\mathfrak{E}$.
(PO7) *If* $a \leqslant b$, *then* $b = a \cup (b^{\perp} \cup a)^{\perp}$.

Given any two elements a, b of a partially ordered set with orthocomplement, write $\alpha \perp b$ whenever $a \leqslant b^{\perp}$. By (PO3) and (PO4) we see that $a \leqslant b^{\perp}$ implies $b \leqslant a^{\perp}$, thus the relation $\perp$ is symmetric (the formulas $a \perp b$ and $b \perp a$ are equivalent). If $a \perp b$ we shall say that a, b are *orthogonal*.

Definition 6.2.3. *A function* μ: $\boldsymbol{B}(Re) \to \mathfrak{E}$ *is an* $\mathfrak{E}$-measure *if and only if the following conditions are satisfied*:
(μ1) *If* $\Delta_1 \cap \Delta_2 = \varnothing$, *then* $\mu(\Delta_1) \perp \mu(\Delta_2)$, *for all* Δ_1, Δ_2 in $\boldsymbol{B}(Re)$.
(μ2) *If* $\Delta_1, \Delta_2, \ldots$ *is an infinite sequence of Borel sets such that* $\Delta_i \cap \Delta_j = \varnothing$ *whenever* $i \neq j$, *then* $\mu(\Delta_1 \cup \Delta_2 \cup \ldots) = \mu(\Delta_1) \cup \cup \mu(\Delta_2) \cup \ldots$
(μ3) $\mu(\varnothing) = 0_{\mathfrak{E}}$.
(μ4) $\mu(Re) = 1_{\mathfrak{E}}$.

Theorem 6.2.4. *Let* $\mathfrak{a} = (F_1, \ldots, F_n)$ *be an elementary system. Denote by* $\mathcal{O}$ *the set of all variables* F *such that the values of* F *are definable in terms of values of* $F_1, \ldots, F_n$ *in* $\{\mathfrak{a}\}$. *Let*

$$P: \mathcal{O} \times St(\mathfrak{a}) \times \boldsymbol{B}(Re) \to \{0, 1\}$$

be defined by the condition

(P) $P(F, \xi, \Delta) = 1$ *if and only if* $F(t) \in \Delta$ *for some* $t \in T(\mathfrak{a})$ *such that* $\xi = \mathfrak{a}(t)$, *or else* $P(F, \xi, \Delta) = 0$.

The triple $(\mathcal{O}, St(\mathfrak{a}), P)$ *is an abstract physical system in the sense of Definition* 6.2.1.

Proof. Before we start proving the theorem let us notice that if $F \in \mathcal{O}$, then for all t, t' in $T(\mathfrak{a})$ $F(t) = F(t')$, whenever $\mathfrak{a}(t) = \mathfrak{a}(t')$, and hence we can treat F as a function defined on $St(\mathfrak{a})$ by the condition: for each ξ in $St(\mathfrak{a})$, t in $T(\mathfrak{a})$

$$\text{(F)} \qquad F(\xi) = F(t) \text{ if and only if } \xi = \mathfrak{a}(t).$$

On the other hand, if $F: St(\mathfrak{a}) \to Re$, then (F) serves as the definition of the function $F: T(\mathfrak{a}) \to Re$ being an element of $\mathcal{O}$.

We have to check one by one whether the axioms by means of which the notion a physical system is defined are satisfied.

Axiom 1 is immediately seen to be valid. The same can be said about Axiom 2.

Indeed, if $P(E, \xi, \Delta_1 \cup \Delta_2 \cup \ldots) = 1$, then $P(E, \xi, \Delta_i) = 1$ for exactly one Δ_i since $\Delta_1, \Delta_2, \ldots$ are pairwise disjoint. Thus $\sum_{i=1} P(E, \xi, \Delta_i) = 1$. And vice versa. If the latter identity holds true, then $P(E, \xi, \Delta_i) = 1$ for exactly one Δ_i and hence $P(E, \xi, \Delta_1 \cup \Delta_2 \cup \ldots) = 1$.

The arguments which allow to establish Axioms 3 and 4 are also trivial. If $P(E, \xi, \Delta) = P(E', \xi, \Delta)$ for all ξ and all Δ, then for each $t \in T(\mathfrak{a})$, $E(t) = E'(t)$, i.e. $E = E'$. To prove that $\xi = \xi'$, if $P(E, \xi, \Delta) = P(E, \xi', \Delta)$ for all E and Δ, select any $t \in T(\mathfrak{a})$ such that $\xi = \mathfrak{a}(t)$ and define $\Delta_1, \ldots, \Delta_n$ by the conditions: $\Delta_i = [x_i, x_i]$, where $x_i = F_i(t)$, $i = 1, \ldots, n$. Let $\xi = \mathfrak{a}(t')$. By the assumptions of Axiom 4, we have $F_i(t') = F_i(t)$, $i = 1, \ldots, n$, i.e. $\mathfrak{a}(t) = \mathfrak{a}(t')$ or equivalently $\xi = \xi'$.

Let $E_1, E_2, \ldots$ and $\Delta_1, \Delta_2, \ldots$ be sequences satisfying the assumptions of Axiom 5. Put

$$s(\xi) = \sum_{i=1} P(E_i, \xi, \Delta_i).$$

For each ξ, $s(\xi)$ equals either 1 or 0. Select in an arbitrary way two real numbers x_1, x_0 and define the function $E_s: S \to Re$ and thus at the same time its counterpart $E_s: T(\mathfrak{a}) \to Re$ (cf. (F)) by the condition

$$E_s(\xi) = \begin{cases} x_1, & \text{if} \quad s(\xi) = 1 \\ x_0 & \text{otherwise.} \end{cases}$$

We obviously have

$$P(E_s, \xi, [x_1, x_1]) = \sum_{i=1} P(E_i, \xi, \Delta_i)$$

for all ξ.

Clearly E_s is in $\mathcal{C}$ and thus we have established Axiom 5.

Assume now that μ is an $\mathfrak{C}$-measure. For each Q in $\boldsymbol{B}(Re)$ select an arbitrary E_Q in $\mathcal{C}$ and Δ_Q in $\boldsymbol{B}(Re)$ such that

$$\mu(Q) = |(E_Q, \Delta_Q)|.$$

Put:

(E_μ) $E_\mu(\xi) \in Q$ if and only if for all ξ, $E_\mu(\xi') \in Q$ if and only if $E_Q(\xi') \in \Delta_Q$, and $E_Q(\xi) \in \Delta_Q$.

We have to prove that E_μ is well defined. Suppose first that $Q \cap Q' = \emptyset$, and for some ξ, $E_Q(\xi) \in \Delta_Q$ and $E_{Q'}(\xi) \in \Delta_{Q'}$. But in order to have simultaneously $E_\mu(\xi) \in Q$ and $E_\mu(\xi) \in Q'$ which would prove inconsistency of the conditions imposed on E_μ, we must have $E_Q(\xi') \in \Delta_Q$ if and only if $E_{Q'}(\xi') \in \Delta_{Q'}$ which yields

(1) $$|(E_Q, \Delta_Q)| = |(E_{Q'}, \Delta_{Q'})|$$

But this is impossible since $Q \cap Q' = \emptyset$ implies that

(2) $$\mu(Q) \perp \mu(Q'),$$

which contradicts (1).

The conditions imposed on E_μ are not contradictory then. Still we have to prove that E_μ is defined in a unique way, i.e. for each ξ, (E_μ) determines exactly one x such that $E_\mu(\xi) = x$.

We clearly have $E_\mu(\xi) \in Re$. On the other hand, if for any Q, $E_\mu(\xi) \in Q$ and

$$Q = Q \cup Q',$$

where $Q \cap Q' = \emptyset$, then either $E_\mu(\xi) \in Q$ or $E_\mu(\xi) \in Q'$, and clearly (E_μ) determines which of these two possibilities actually takes place. In this way, taking Re as the initial element (i.e. $Re = Q_1$), we can form a sequence

$$Q_1, Q_2, \ldots$$

of intervals such that for each i

(i) $Q_i \subset Q_{i-1}$, when $i \geqslant 2$,
(ii) $E_\mu(\xi) \in Q_i$,
(iii) If $E_\mu(\xi) \neq y$, then for some Q_i, $y \in Q_i$.

We have to prove that the intersection $\bigcap Q_i$ is not empty. Clearly then

$$E_\mu(\xi) = \bigcap Q_i.$$

Suppose that $\bigcap Q_i$ is empty and consider the sequence

$$Q'_1, Q'_2, \ldots$$

where $Q'_i = Q_i - Q_{i+1}$. We obviously have

$$E_\mu(\xi) \notin Q'_i \tag{3}$$

for all $i \geqslant 1$
At the same time we have

$$Q'_i \cap Q'_j = \varnothing \tag{4}$$

for all $i \neq j$, and

$$Q'_i = Re. \tag{5}$$

Consider any Q'_i. Obviously $\mu(Q'_i) = |(E_{Q'_i}, \Delta_{Q'_i})|$ and by (3) and (E_μ) we obtain

$$E_{Q'_i}(\xi) \notin \Delta_{Q'_i}. \tag{6}$$

By (5) and (E_μ) we have

$$\mu(\bigcup Q'_i) = \mu(Re) = |(E_{Re}, \Delta_{Re})|.$$

On the other hand (4), (5), (6) and condition (μ2) imposed on μ (cf. Definition 6.5.3) yields

$$E_{Re}(\xi) \notin \Delta_{Re}.$$

But this is impossible since, by (μ4), $\mu(Re) = 1$.

BIBLIOGRAPHY

Ashby, W. R.

(1956) *An Introduction to Cybernetics*, London.

(1960) *Design for a Brain*, London.

Ajdukiewicz, K.

(1958) 'Le Probléme du Fondement des Propositions Analytiques', *Studia Logica*, **8**, 259–272.

Bridgman, P. W.

(1927) *The Logic of Modern Physics*, New York.

(1950) *Reflections of a Physicist*, New York.

Bunge, M.

(1973) *Philosophy of Physics*, Dordrecht.

Beth, E. W.

(1949) 'Towards an Up-to-date Philosophy of the Natural Science', *Methods*, **1**, 178–185.

(1960) 'Semantics of Physical Theories', *Synthese*, **12**, 172–175.

Burks, A. W.

(1951) 'The Logic of Causal Propositions', *Mind*, **60**, 363–382.

Carnap, R.

(1942) *Introduction to Semantics*, Harward.

(1952) 'Meaning Postulates', *Philosophical Studies*, **3**, 65–73.

(1955) *Foundations of Logic and Mathematics. International Encyclopaedia of Unified Science*, vol. 1, no. 3, Chicago.

(1958a) 'Beobachtungssprache theoretische Sprache', *Dialectics*, **12**, 34–42.

(1958b) *Introduction to Symbolic Logic and its Applications*, New York.

(1958c) *Meaning and Necessity*, Chicago.

(1966) *Philosophical Foundations of Physics*, London.

Campbell, N. R.

(1920) *Physics: the Elements*, Cambridge, Reprinted as *Foundations of Science: the Philosophy of Theory and Experiment*, Dover, New York, 1957.

(1928) *An Account of the Principles of Measurement and Calculation*, London.

Cohen, M. R., and Nagel, E.

(1934) *An Introduction to Logic and Scientific Method*, New York.

Dalla Chiara Scabia, M. L., Toraldo di Francia, G.

(1973) 'A Logical Analysis of Physical Theories', *Rivista del Nuovo Cimento*, Serie 2, Bol. 2, 1–20.

Domotor, Z.

(1972) 'Causal Models and Space Time Geometries', *Synthese*, **24**, 5–57.

Duhem, P.
(1914) *La Théorie Physique*, Paris.
Feynman, R. P., Leighton, R. B., Sands, M. I.
(1966) *The Feynman Lectures on Physics*, Reading.
Fine, K.
(1975) 'Vagueness, Truth and Logic', *Synthese*, **30**, 265–300.
Giles, R.
(1974) 'A Non-classical Logic for Physics', *Studia Logica* **33**, 4, 397–415.
(1975) *Formal Languages and the Foundations of Physics*, Mimeographed.
Good, I. J.
(1961) 'A Causal Calculus (I)', *The British Journal for the Philosophy of Science*, **11**.
(1962) 'A Causal Calculus (II)', *The British Journal for the Philosophy of Science*, **12**.
Grzegorczyk, A.
(1974) *An Outline of Mathematical Logic*, Warsaw.
Gudder
(1967) 'Systems of Observables in Axiomatic Quantum Mechanics', *Journal Mathematics Phys.*, **8**.
Kalman, R. E., Falb, P. L., and Arbib, M. A.
(1969) *Topics in Mathematical System Theory*, New York.
Kochen, S. and Specker, E. P.
(1965) 'Logical Structures Arising in Quantum Theory', in: J. Addison, L. Henkin, A. Tarski (Eds.), *The Theory of Models*, Amsterdam.
Kokoszyńska, M.
(1963) 'Über den absoluten Wahrsheitsbegriff und einige andere semantische Begriffe', *Erkenntnis*, **6**, 143–165.
Krantz, D. H., Luce, R. D., Suppes, P., Tverski, A.
(1971) *Foundations of Measurement*, vol. I., New York and London.
Kuhn, T. S.
(1962) *The Structure of Scientific Revolutions*, Chicago.
Landan, L. D., Lifshits, E. M.
(1969) *Mechanics. Electrodynamics* (in Russian), Moscow.
Latzer, R. W.
(1974) 'Errors in the no Hidden Variable Proof of Kochen and Specker', *Syntese*, **29**, 331–372.
Łoś, J.
(1963) 'Remark on Foundations of Probability, Semantical Interpretation of the Probability of Formulas', in *Proceedings of the International Congress of Mathematicians* 1962, Stockholm.
(1963) 'Semantic Representation of the Probability of Formulas in Formalized Theories', *Studia Logica*, **14**, 183–195.
Mackey, G.
(1963) *Mathematical Foundations of Quantum Theory Mechanics*, New York.

Mączyński, M. J.
(1968) 'A Remark on Mackey's Axiom System for Quantum Mechanics', *Bull. Acad. Polon. Sci. Ser. Math., Astr. et Phys.*, **15**, 583–587.
McKinsey, I. C., Sugar, A. C., Suppes, P.
(1953) 'Axiomatic Foundations of Classical Particle Mechanics', *Journal of Rational Mechanics and Analysis*, **2.**
Mehlberg, H.
(1958) *Reach of Science*, Toronto.
Montague, R.
(1962) 'Deterministic Theories' in E. Wahsburne (ed.) *Decisions, Values and Groups*, Oxford (reprinted in *Formal Philosophy*, New Haven and London (1974).
Nagel, E.
(1953) *The Structure of Science*, New York.
Nowak, S.
(1972) 'Inductive Inconsistencies and Conditional Laws of Science', *Syntheses*, **23.**
Popper, K.
(1963) *Conjectures and Refutations*, London.
(1972) *Objective Knowledge*, Oxford, London.
Przełęcki, M.
(1969) *The Logic of Empirical Theories*, London.
(1976) 'Fuzziness as Multiplicity', *Erkenntnis*, **10**, 371–380.
Przełęcki, M., Szaniawski, K., and Wójcicki, R. (eds)
(1977) *Formal Methods in the Methodology of Empirical Sciences*, Dordrecht.
Przełęcki, M. and Wójcicki, R.
(1969) 'The Problem of Analyticity', *Synthese*, **19**, 49–74.
(1971) 'Inessential Parts of Extensions of First-Order-Theories', *Studia Logica*, **28**, 83–99.
Rabin, M. O., and Scott, D.
(1959) 'Finite Automata and their Decision Problems', *IBM Journal of Research and Development*, **3**, reprinted in E. F. Moore (Ed.), *Sequential Machines*, Reading, Massachusetts: Addison-Wesley, 1964.
Ramsey, F. P.
(1931) *The Foundations of Mathematics and Other Logical Essays*, London.
Rantala, V.
(1975) *Prediction and Identifiability*, Mimeographed.
Rasiowa, H., Sikorski, R.
(1963) *The Mathematics of Metamathematics*, Warszawa.
Reese, T. W.
(1943) 'The Application of the Theory of Physical Measurement to the Measurement of Psychological Magnitudes, with three Examples', *Psychol. Monogr.*, **55.**

Robinson, A.
(1963) *Introduction to Model Theory and to the Metamathematics of Algebra*. Amsterdam.

Scheibe, E.
(1973) 'The Approximative Explanation and the Development of Physics', P. Suppes *et al.*, eds., *Logic Methodology and Philosophy of Science* IV.

Schoenfield, J. R.
(1967) *Mathematical Logic*, Reading.

Scott, D. and Suppes, P.
(1958) 'Foundational Aspects of Theories of Measurement', *Journal of Symbolic Logic*, **23**, 113–128.

Simon, H. A.
(1954) 'Spurious Correlations: A Causal Interpretation', *Journal of American Statistical Association*, **19**.

Smith, J. M.
(1968) *Mathematical Ideas in Biology*, Cambridge.

Smith, J. M., and Price, G. R.
(1973) 'The Logic of Animal Conflicts', *Nature*, **246**.

Sneed, J.
(1971) *The Logical Structure of Mathematical Physics*, Dordrecht.

Stegmüller, W.
(1973) *Theorie and Erfahrung*, Berlin–Heidelberg–New York.

Suppe, F.
(1972) 'Theories, their Formulations and the Operational Imperative', *Synthese*, **25**.
(1976) 'Theoretical Laws', in: M. Przełęcki, K. Szaniawski and R. Wójcicki (Eds.), *Formal Methods in the Methodology of Empirical Sciences*, Dordrecht.

Suppes, P.
(1957) *Introduction to Logic*. Princeton.
(1961) 'The Meaning and Uses of Models', in: (Eds.). *The Concept and the Role of Model in Mathematics and Natural and Social Science*, Dordrecht.
(1970) *A Probabilistic Theory of Causality*, Amsterdam.
(1970a) *Set-Theoretical Structures in Science*, forthcoming (mimeographed, Stanford University).
(1974) 'The Measurement of Belief', *The Journal of the Royal Statistical Society*, Series B (Methodological), vol. 36, No. 2.

Suszko, R.
(1957) Formal Logic and Some Problems in Epistemology (in Polish), *Myśl Filozoficzna*, **2** and **3**.
(1968) 'Formal Logic and the Evolution of Knowledge', in: I. Lakatos (ed.) *Problems in the Philosophy of Science*, Amsterdam.

Tarski, A.

(1930) 'Über einige fundamentale Begriffe der Metamathematik', *Comptes Rendus des séances de la société des Sciences et des Letteres de Varsovie*, **23**, 22–29. Reprinted as 'On some fundamental concepts of metamathematics' in *Logic, Semantics, Mathematics*, Oxford, 1956.

(1933) 'Pojęcie prawdy w językach nauk dedukcyjnych', *Travaux de la Société des Sciences et des Letteres de Varsovie, Classe III, Sciences mathématiques et physiques*. Reprinted as 'The Concept of Truth in Formalized Languages' in *Logic, Semantics, Metamathematics*, Oxford.

(1944) 'The Semantic Conception of Truth', *Philosophy and Phenomenological Research*, **4** (reprinted in: H. Feigl, W. Sellars (Eds.), *Readings in Philosophical Analysis*, New York.).

(1956) *Logic, Semantics, Metamathematics*, Oxford.

Tuomela, R.

(1973) *Theoretical Concepts*, Wien–New York.

Van Fraassen, B. C.

(1970) 'On the extension of Beth's semantics of physical theories', *Philosophy of Science*, **37**, 325–339.

(1975) 'A Formal Approach to the Philosophy of Science', in R. Colodny (ed.), *Paradigms and Paradoxes*, Pittsburgh.

von Neumann, J., Morgenstein, O.

(1953) *Theory of Games and Economic Behaviour*, Princeton.

Weyl, H.

(1927) 'Philosophie der Mathematik und der Naturwissenschaften'. *Handbuch der Philosophie*, München–Leipzig.

Williams, P. M.

(1974) 'Certain Classes of Models for Empirical Theories', *Studia Logica*, **33**, 73–90.

Wójcicki, R.

(1969) 'Semantyczne pojęcie prawdy w metodologii nauk empirycznych', *Studia Filozoficzne*, **3**, 33–48 English translation 'Semantic Conception of Truth in Methodology, of Empirical Sciences', *Dialectic and Humanism*, **1** (1974), 103–115.

(1975) 'Deterministic Systems', *Erkenntnis*, **1**, 219–227.

(1975a) 'The factual content of empirical theories', in J. Hintikka, R. Carnap (Ed.), *Logical Empiricist*, Dordrecht 1975, 95–122.

INDEX OF SYMBOLS

(some of the symbols are listed along with the typical context in which they appear)

INDEX OF NAMES

SUBJECT INDEX